Advances in Cotton Physiology

Advances in Cotton Physiology

S K Thind
Associate Professor
Department of Botany
Punjab Agricultural University,
Ludhiana, Punjab (India)

2007

SSPH SATISH SERIAL PUBLISHING HOUSE
403, Express Tower, Commercial Complex, Azadpur, Delhi - 110033 (INDIA)
Phone : 011-27672852 Fax : 91-11-27672046
E-mail : info@satishserial.com, hkjain1975@yahoo.com
Website : www.satishserial.com

Published by :

SATISH SERIAL PUBLISHING HOUSE
403, Express Tower, Commercial Complex, Azadpur, Delhi - 110033 (INDIA)
Phone : 011-27672852 Fax : 91-11-27672046
E-mail : info@satishserial.com, hkjain1975@yahoo.com

ISBN 81-89304-35-6

Printed at
SALASAR IMAGING SYSTEMS, DELHI-35

Contents

1

Historical Background and Physiological Terms

About 30 species of cotton have been identified around the world, but only four are cultivated commercially. Most commercial cottons of the world, however, are predominantly highly selected strains of *Gossypium hirsutum*. Cotton *(Gossypium sp.)* is one of the most important cash crops of the world. Cotton belongs to the family *Malvaceae*. Further, cultivated cotton are *Gossypium arboreum, Gossypium hirsutum, Gossypium herbaceum* and *Gossypium barabadense*. It has multiple uses in the textile as well as other industries.

One of the oldest evidence of cultivation of cotton for thread and cloth was reported in region of Western India and Pakistan. The metallurgical analysis of a copper bead from a Neolithic burial (6th millennium BC) at Mehrgarh, Pakistan, allowed the recovery of several threads, preserved by mineralisation. They were characterised according to new procedure, combining the use of a reflected-light microscope and a scanning electron microscope, and identified as cotton *(Gossypium* sp.). The Mehrgarh fibres constitute the earliest known example of cotton in the Old World and put the date of the first use of this textile plant back by more than a millennium. Even though it is not possible to ascertain that the fibres came from an already domesticated species, the evidence suggests an early origin, possibly in the Kachi Plain, of one of the Old World cotton.

Commercial upland cotton, *Gossypium hirsutum* L., is native to Mexico and Central America where it can be found growing in the wild as a bushy shrub or tree, perhaps twenty five feet tall. Cotton plant with an indeterminate growth habit, has been adapted for commercial production as an annual crop. Although often grown in humid or irrigated areas, it is classified as a xerophyte (i.e., a plant adapted to live in an arid climate). Variation in growth habit among cotton genotypes has led to the grouping of cultivars into "determinate" and "indeterminate" classes. Although all cottons are botanically classified as indeterminate, these variety groupings refer to the strong selection pressure that commercial breeders have placed on cotton to produce an early crop compared to wild cotton plants. "Determinate" cultivars are generally classed as early or short-season varieties, and *"indeterminate"* cultivars are referred to as late or full-season varieties.

Evolution and Genetic Diversity

Under domestication pressures, *Gossypium hirsutum* and *G. barabadense* evolved from a sprawling shrubs or trees with strict short day photoperiodic flowering, short seed coat fibers and remarkable seed dormancy into compact, annualized, day-neutral crops, with negligible seed dormancy and astonishingly long seed coat fibers. Since, modern cultivars are interfertile with 'wild' and primitive taxa thus represent various stages of domestication. The genetic basis for these evolutionary changes can be uncovered through molecular genetics and comparative genomic approaches. PCR-based molecular marker such as AFLPs, microsatellites and CAPS have been used to uncover the evolutionary and genetic relationships of a variety with wild, primitive and domesticated *G. hirsutum* taxa. These works are useful both for phytogenetic and phonetic studies as well as genome mapping. Microsatellite and CAPS contain imbedded sequence information that can be useful for the characterization of evolutionarily and economically important traits in *Gossypium,* such as vegetative morphology, fiber development and the timing of flowering. The tools derived from genetic and molecular studies of flowering time in model plant have been used to understand photoperiod sensitivity in the short-day plant *Gossypium hirsutum.* This work can be used for the improvement of cultivated cottons for enhanced insect and pathogen resistance and improved fiber yield as well as quality.

Various terms used for recording morphophysiological observations are :

Abscission

Loss of a plant structure from an abscission zone (for example, when a plant loses a leaf).

Abscission zone

Area at the base of a leaf petiole or peduncle of a square or boll where the plant structure separates during abscission.

Anthesis

Developmental stage of a flower when fertilization takes place, generally associated with white flowers.

Average internode length

Height (length) of the main stem divided by the number of main-stem nodes.

Axil

The upper angle between the petiole of a leaf and the stem from which it grows.

Boll count

Average number of bolls per sampled unit.

Boll load

Cotton plant's capacity to retain and develop fruit, within limits of its genetic constitution and the prevailing environmental conditions.

Boll opener

A chemical, such as ethephon or PrepTM, that increases the rate of boll opening at crop maturity.

Canopy

The covering formed by the leaves of the plant population in a crop.

Compensation

The cotton plant's response to fruit loss in terms of new growth, added development, or adjusted fruit retention.

Cotyledonary node

Opposite nodes at the base of the plant where cotyledons (seed leaves) attach to the plant.

Cracked boll

First visual sign of boll opening, when the carpel walls separate along sutures.

Crop termination

Ceasing active crop growth to prepare for harvest.

Cutout

Termination or extended lapse in terminal growth, due to the development of the boll load sink and the resulting demand for available nutrient and photosynthate resources.

(a) Physiological cutout

Crop development stage characterized by an average number of nodes after white flower (NAWF) of 5.

(b) Seasonal cutout

When the last effective flower date is determined by end-of-season weather restraints rather than crop maturity.

(c) Premature cutout

Early cutout associated with excessive stress.

Damaged (aborted) terminals

Damaged main-stem tissue in the plant apex, causes a loss of apical dominance.

A method used to assist with end-of-season management decisions based on the development of the last effective boll population sequenced by growing degree days accumulation. Research has shown that 350 heat units are required after physiological or seasonal cutout.

Defoliant

Any of a variety of chemicals that induce leaf drop when sprayed on the cotton plant.

Defoliation

Removal of the plant's or crop's leaves by a defoliant, crop development, weather, mechanical means, or certain insects and diseases.

Earliness

An imprecise term used to represent the rapid development or maturation of the harvestable cotton crop relative to the available growing season.

Effective fruiting period

The time between first square and cutout.

First flower

Stage of crop development at which about one-half of the plants in a population have developed at least one white flower.

First fruiting branch

First formed sympodial branch on the main stem.

First fruiting node

Lowest main-stem node above the cotyledonary node from which a sympodial branch develops. Usually occurs at main-stem nodes 5 to 9.

First-position bolls

Bolls on sympodial branches at the nodal position nearest to the main stem.

First square

The first fruiting bud to appear on a cotton plant. Also the stage of growth when approximately 50% of the plants have a visible square.

Flowering interval

The time in days or heat units between the appearance of white flowers either at adjacent fruiting positions along the same sympodium (horizontal flowering interval) or at the same fruiting position on the next higher sympodial branch (vertical flowering interval).

Fruit retention

Presence of a square or boll at a fruiting position.

Fruit shedding

Physiologically induced separation of a square or boll from a plant, usually induced by some damage to squares/bolls or stress on the plant.

Fruiting node

Main-stem nodes that produce sympodia or fruiting branches.

Fruiting position

A specific nodal position on a sympodium where a reproductive structure may be produced. Positions may be sequentially numbered from the site nearest the main stem. A fruiting position may have a scar (where fruit is aborted), a square, a flower, or a boll.

Growing degree days or DD6O's

A concept based on a temperature threshold (usually15.5 degrees C for cotton) above which the crop grows and below which little or no development occurs.

Harvest (picking) capacity

The number of acres of cotton that can be picked per day.

Harvestable boll

A mature or immature boll that will open before the end of the growing season to permit mechanical harvest.

Height-to-node ratio

Average internode length determined by dividing the total plant height in inches by the total number of main-stem nodes.

Internode

The stem section between two nodes, that is, the space between two successive true leaves on the main stem or a branch.

Last effective flower/boll population

The latest developing population of fruit that has a high probability of being retained and developing into bolls of adequate size and fiber properties.

Last possible flower/boll

The latest developing individual flower/boll that will develop into a harvestable boll.

Late crop

A subjective term that refers to bolls produced on the upper part of the plant canopy usually late in the season.

Main stem

The central axis of the plant consisting of a terminal meristem and a series of internodes with one main-stem leaf at each node.

Main-stem node

The part of the stem at which a main-stem leaf is attached.

Maturity

Describes the completion of natural growth and development.

a) Crop maturity

Crop maturity is related to a field population of plants that has developed to the point that no additional inputs are required.

b) Boll maturity

A mature boll is one that has sufficient nutrition to open normally if the subtending leaf is removed.

Node

A point of attachment of plant structures (leaves or fruit) to the main stem or branches of a plant.

Nodes above first square (NAFS)

A measure of the number of main-stem nodes above the first fruiting node or first square.

Nodes above white flower (NAWF)

A measure of the number of main-stem nodes above the uppermost white flower in the first fruiting position.

Open boll

A boll in which the carpel sutures have abscised and allowed the seedcotton to become exposed and dry.

Plant monitoring

One of a series of measurements on the plant or crop, usually made sequentially during the growing season.

Plant population

An estimate of the number of plants per unit area.

Regrowth

The resumption or continuation of growth that may occur after applying a harvest-aid chemical.

Sample site

Location in the field where samples are taken. Generally, four or more sample sites will be randomly selected for each field.

Sampling

Process of selecting and monitoring a subset of the field population of plants to estimate the status of the entire field.

Scar

A mark left on the stem or branch where a leaf square or flower abscised.

Shed/shedding

Separation of a leaf, square, or boll from a plant, usually induced by some stress.

Squaring nodes

Collective term for NAFS (prior to first flower) and NAWF (after first flower).

Stand (effective stand)

The number of plants per area, e.g., plants per acre.

Stratum (Strata)

Divisions of a field that enhance sampling designs. For representative sampling, each field may be divided into four or more strata.

Sympodium (sympodia)

A segmented fruiting branch where flowers and resulting bolls grow.

Terminal

Usually refers to the growing point in the plant apex.

Terminal node

Uppermost main-stem node from which the main-stem leaf is unfurled.

Unfurled leaf

A young developing leaf that has expanded such that the unfolding edges no longer touch each other.

Vegetative node

Main-stem nodes between the cotyledonary node and the first fruiting node.

Vigor index

A measure of early-season growth of cotton plants, typically by evaluating plant height over time or plant height in relation to the number of main-stem nodes, e.g., height-to-node ratio.

White flower

A cotton flower with white petals on the day of pollination. Following pollination, the flower petals turn pink then red.

□□□

2

Botanical Description

Classification :Division : Magnoliophyta, Subdivision, Magnoliopsida, Order : Malvales, Family Malvaceae

Common names

algodao, algodón, baumwolle, bomuld, coton, cotton, från amerika, kapas, katoen, mianhua

It is an annual subshrub, up to 1.5 m tall; branches of two kinds: vegetative and fruiting. It is the most important of the cotton-yielding plants, providing the bulk of commercial cottons. Seeds yield a semi-drying oil and the processing residue, cottonseed cake or meal, is an important protein concentrate for livestock.

Roots: The well-developed taproot with numerous laterals grows downward into moist soil at a rate of about 2.5 cm per day. Cotton seedlings establish a deep root system very early. The tap root has a thick cortex and wood more or less has a quadrangled structure with four well defined ridges. Sometimes the branches that arise from the tap root are irregularly grouped into four rows, but only three are recognized because of twisting and other irregularities.

The tap root tapers rapidly after just 30 cm below the ground level. In a well grown tap root, at about 60 cm depth, it is no more than about 1.5 mm in diameter. Then if conditions are favourable, it may grow downwards to a distance of more than a meter without any change in the diameter. From the tap root, many laterals arise and radiate outwards with a slightly downward orientation. These laterals branch and rebranch and form a network of fine thread like roots.The root system of young cotton plant develop rapidly. In a 7 day old seedling, the total length of the roots including tap root and lateral may be more than 20 cm under favourable conditions. Similarly, in a 14 day old seedlings it may be 2 meters long.

Stems: Vegetative branches have a development pattern almost similar to that of the main stem. They may often grow upright and they continue to develop from an apical bud.Vegetative branch may vary from one to many depending upon various factors. They contain only the leaves and are generally straight in orientation. Fruiting branches are those branches which bear the reproductive parts of flowers, fruits etc. A boll always has a subtending leaf attached to it in a normal plant, which contributes maximum of nutrients to the developing boll. In order to have a good yield, more number of fruiting branches are essential.

Leaves: Cotton leaves have some distinctive features. They have two pulvini at each end of the petiole which allow a wide arc of movement of the leaf blade to follow the movement of the sun. They have leaf nectaries on the veins. Leaves also have pigment glands at random.Two to three times more stomata are present on the lower surface of leaves.The first true leaf of the primary axis does not have lobes .At the base of each cotton leaf petiole there are two buds, one called the true axillary bud which continues the vegetative growth, and the other one the extra axillary bud which produces the fruiting branch. The leaves are petioled, stipulate, and have three, five, or seven lobes. Leaves and stems are usually covered with fine hairs. The flowers may appear arranged on alternate sides of the fruiting branch. The young flower buds are called squares which are enclosed within three relatively large leaf-like bracts.

Flowers: The first formed buds are small, green and also pyramidal in structure called 'square'and then the flower forms.. 25-30 days are required for a square to form into a flower after which the boll formation and maturation takes place. The flowers are arranged on alternate sides of the sympodial branches. There are three large leaf-like bracts at the base of the flower. Above this is the true calyx which consists of unequally lobed sepals. The corolla consists of 5 petals which are generally yellow in colour if the flower had bloomed on the same day. A flower is produced regularly at the first node of the next higher fruiting branch .Flowers appear on the successive positions of the same fruiting branch at intervals of about 6 days. The number of squares produced is directly related to the number of fruiting branches and number of fruiting positions. Peak flowering occurs at about a month after first bloom, levels of for 2-3 weeks and then declines. The staminal column bears ten or more or less double rows of stamens and the pistil is made up of three or five carpels. The stigma tip extends above the anthers.

Pollination: Cotton is readily autogamous, while out crossing is less. The extent of cross-pollination varies with the population of bees and other insects. Metaxenia, effect of foreign pollen sometimes reduces the length of the fibers.

Fruits: Boll size depends on the soil characteristics, water management and population density as well as the structure of the plant. All these contribute to boll size and number.The number of loculi varies from 3-5 per boll. A healthy cotton plant produces more flowers and fruits than can possibly mature. Only about 25-35% of the fruits formed finally mature if the environmental conditions are favourable for it. This helps in offsetting the loss of yield due to temporary unfavourable conditions which occur during the growing season.

Seeds: Seeds in relatively early formed bolls are nourished through the high photosynthetic capability of a relatively young plant whereas seeds from the later formed bolls are nourished from an old and aging plant.The production of seed crop is dependent on the leaf canopy. Anything that affects the development of a the plant canopy will also affect the seed quality. The quality of the seed is maximum if it matures on the plant. Normally there are about 27 to 45 seeds per healthy boll. The seeds are usually ovoid in shape. Most of the upland cotton varieties have dark brown seeds covered with fuzz. Large bolled varieties tend to have large seeds. The seed coat is a tough leathery hull that constitutes about 25 to 30 percent of the weight of the seed. The oil content of the hulled kernel varies from 32 to 37 per cent.

Lint or Fiber : Cotton fibers are slender single-cell hairs that grow out from certain epidermal cells of the cottonseed. Their growth starts on the ovules about the time the flower opens. It first emerges as a slight swelling of the outer wall of the epidermal cell on the day of flowering and thereafter, every day it begins to elongate. Not all cells of the epidermis develop into fibres, some form normal fibres, others develop into short thick fibres, called "fuzz", while the rest remain as mere epidermal cells. The fiber lengthens rapidly and attains full length in about 18 days when the seed has attained full length. For an additional 3-4 weeks, the cell walls of the fiber thickens On the first day of flower opening, most of the differentiated fibres, appearing as small protrusions, are seen at chalazal end of the seed, while on the second, third and subsequent days further protrusions appear, indicating that the pattern of these protrusions spreads towards the pointed or micropylar end of the seed. It has been suggested that the fibre cells that sprout from the first to the fifth day develop into commercial lint fibres, while those which appear later develop into fuzzy fibres which form a dense entangled mat that adheres closely to the seed surfaces.

Thickening occurs by the growth of two additional spiral rings each day on the inner surface of the cell wall. The fiber is cylindrical before the boll opens, but collapses and becomes more or less flattened and twisted (with convolutions) with the opening of the boll. The fiber has a very thin wall and is very weak if the climatic conditions during fiber maturation is not favourable. The length of the fiber may be 1000 - 3000 times its thickness. The thickness may be about 0.0006 to 0.0008 inches while its length may be from 0.25 inch to 2 inches.

The fiber is the seed hair. It is a hyperelongated single cell arising from the protodermal cells of the outer integument layer of the seed coat. These cells respond to the variations and fluctuations in the environment during their growth stage. Thus, the fibres on a single seed constitute continua of fibre length, shape, cell-wall thickness, and physical maturity. The lint and fuzz hairs develop as thin tubular structures to their maximum lengths, then the walls of each thicken from within through the formation of consecutive laters of cellulose on the inner wall. At this active growing stage the hairs (as living cells) are filled with protoplasm, with a nucleus.After the fibers reach their maximum length the walls are thickened by the deposition of consecutive layers of cellulose on their inner surface in a spiral fashion. Secondary thickening occurs just before the boll opens. The degree of thickening and the angle of the spirals affect the fiber maturity and strength. Seeds of some cultivars require a 2-3 month period of dormancy. Seeds lose viability quickly under moist conditions.

Growth and Development

Cotton is a crop of warm plains, grown commercially from sea level to 1200 m, with some perennial forms found at 1800 m. Traditionally cotton has been restricted to altitudes below 1250 m. However it could be successfully cultivated at higher altitudes provided the temperature does not fall too rapidly at the end of the season with an annual temperature of 7.0

to 27.8°C. Good sunlight and warmth is required. So cotton growing regions are generally confined to warmer regions. It is sensitive in any stage to frost so cotton limits are set by the early and late frosts.

A long-season plant, cotton requires a minimum of 180-200 frost-free days of uniformly high temperatures, the optimum temperature required by cotton is around 25°C. Generally three climatic essentials are freedom from frost for a minimum growing and ripening season, an adequate supply of moisture, and abundant sunshine. The minimum, optimum and maximum temperatures for germination and early growth of cotton are about 60, 93 and 102 degrees F respectively. The most rapid growth and flowering of the plants occur at temperatures of 91 to 97 degrees F.

Upland Cotton is reported to tolerate annual precipitation of 29 to 2780 mm .It requires at least 450-500 mm of water for evapotranspiration, in order to give high yields. Irrigation water is used most efficiently by cotton plants when applied from first boll appearance. Boll maturity may be sometimes delayed due to irrigation. Ideally it needs 2.5mm per day at the beginning of growth, 5 mm per day at flowering, 5.8 mm per day at full bloom, 6 - 7 mm per day at the end of flowering and 4 - 6 mm per day at boll formation.

Radiation: Enough sunlight is needed in the early, boll and ripening stage. Otherwise there wont be uniform bursting of bolls. Full sunlight is critical for proper development .Periods of low light intensity or short day periods of full sunlight impair the fruiting capacity of the plant and increase the vegetative growth depending upon the variety. Short photoperiods markedly decrease plant height, increase the number of nodes and sympodial branches, shorten internodal length.

Soil: Cotton grows best on well drained, fertile soils with a high moisture holding capacity and of unrestricted depth. This is because of its deep root system and long season requirements. The crop may also be cultivated in relatively shallow soils, provided the subsoil does not restrict root penetration. In this case, management of the crop becomes very critical in order to obtain optimum yields. However, in light textured soils, if proper tillage, moisture conservation measures, nutrient and water management aspects are taken care, then it is possible to cultivate in those soils also. In India, cotton is grown usually on black alluvial and red soils. Mixtures of clay and sandy loam containing fair amounts of organic matter and a moderate amount of available nitrogen, phorphorus and potash are best suited for cotton. Soils which are prone to water logging should be avoided to the extent possible. In such cases, special tillage practices which ensures proper drainage of excess water should be ensured. Special attention should also be paid to avoid erosion of top soil from a cotton field in the early stages because, cotton takes around 8 weeks for complete canopy cover over the ground. It can bear pH of 4.5 to 8.4 and Optimum pH lies between 5.2-7.

The soil surrounding the roots from about 1-6 mm is called the rhizosphere. The surface of the root is called the rhizoplane. The cotton field soil environment which includes the roots as well as the soil intimately in contact with the roots, is very dynamic in nature. The root exudates form a very important source of nutrients for the microorganisms which inhabit the soil surround the root.The root exudates contain amino acids, sugars, vitamins, enzymes and other organic acids. Collectively, these compounds act as an excellent media for microbial growth. It is essential to understand the inter relationship between the micro organisms and the rhizosphere though it is extremely complex in nature. This understanding of the relationship can help in exploiting the associative or antagonistic behaviors of different micro organisms vis a vis the cotton plant. Depending upon the proportion of different compounds in the root exudate, they can selectively foster or suppress the soil micro organisms. More than 20 different types of amino acids have been found in the root exudates so far.

Medicinal Uses: Cottonseed and roots have been used in nasal polyps, uterine fibroids and other types of cancer . Gossypol has shown anticancer activity in the tumor systems. Mucilaginous tea of fresh or roasted seeds used for bronchitis, diarrhoea, dysentery, and haemorrhage. Flowers diuretic and emollient, used for hypochondriasis. Leaves steeped in vinegar are applied to the forehead for headache. Often used for abortion; apparently with no serious side effects. The leaves and seeds considered as an emollient and the roots as an emmenagogue. Root decoction used for asthma, diarrhoea, and dysentery.Root bark, devoid of tannin; astringent, used as an emmenagogue, haemostat, lactagogue, oxytocic, and vasoconstrictor. Gossypol is used as a male contraceptive.

3

Morphological Features of Crop

Cotton (*Gossypium* spp.) is a unique and intriguing plant. In nature, it is a woody, perennial, semiarid shrub, often reaching the size of a small tree. No cottons are "true" annuals. Some can grow, fruit, and partially mature within the growing season in the temperate zones. Such cottons are referred to as "annuals". Because cotton developed over time under very dry conditions, it has the capacity to compensate for considerable drought and other stress. Because it is in reality a perennial, its basic response to stress is survival. In contrast, annual plants under stress tend to reproduce. For thousands of years, man has sought to improve cotton by developing new varieties,. through improved fertility and other cultural practices, and by weed, insect, and disease control. However, the fact remains that, man is taking a perennial plant and forcing it to behave as an annual within short-season production. Cotton can be very responsive to management inputs. For example, although the plant is drought tolerant, irrigation during flowering often increases lint yield and fiber quality. In many situations, timeliness of the management practice is of crucial importance. For example, irrigation too soon before flowering or too late after drought stress begins can delay maturity of the fiber and reduce yield.

Growth and Development

Growth and development of the cotton plant follow a regular pattern with formation of roots, leaves, stems, and flowers in an orderly and predictable sequence. Understanding how the cotton plant grows is useful when making management decisions during the season.

Germination

Seed quality is a primary factor influencing germination of cottonseed. The most viable, vigorous planting seed are those that have fully developed embryos. Many things influence the ability of cotton seed to germinate, e.g. these have remained reasonably dry in the field and, after ginning, have been stored in a relatively cool, dry place until planted. In general, the dryer the seed and cooler the storage temperature, the better the chance that the seed will retain their viability and vigour. Drying below 10 per cent moisture is essential to safe storage on the farm, even for a short period.

The reserve food supply of the cotton seed is stored in its two cotyledons. The oxygen, moisture, and temperature requirements for cotton seed are higher than for many other crop plants. Cotton seed is very rich in oil and protein. These reserve foods require large amounts of oxygen for conversion to simple foods needed in plant growth, much more than that needed for the conversion of starch in seed such as corn and wheat. This high oxygen requirement for germinating cotton seed points to the need for maintaining good soil aeration, especially during germination. Optimum germination of cotton seed occurs at a soil temperature of 33.9°C. The minimum soil temperature for both germination and early seedling growth averages about 15.56°C, and the maximum about 38.9°C.

Under favourable conditions, when a cottonseed germinates in the soil, the embryo enlarges very rapidly. The primary root develops from the radicle, forces its way through the seedcoat at the pointed (micropylar) end, and grows downward into the soil (Fig. 3.1). The primary root may reach a depth of 20.32 to 30.5 cms prior to seedling emergence. The hypocotyl, the part of the embryo's stem between the primary root and the attachment point of the seedling leaves, elongates rapidly and arches into a "crook" as it pushes its way upward through the soil. The arch protects the tender tissues of the growing point (apex) from abrasion with particles in the soil. The seedcoat is usually left behind in the soil during emergence (Fig. 3.2). The cotton seedling is weak and delicate. Cottonseed should be planted from 2.54 to 5.08 cms deep, depending on soil texture and friability at planting.

Roots

During early vegetative growth, development of a strong root system is the plant's top priority. The taproot penetrates through the soil rapidly and may reach a depth of 25.4 cms by the time the cotyledons unfold. Root may grow at a rate of 1.27 to 5.1 cms per day during the early vegetative stage, permiting early roots to be twice as long as the plant height.

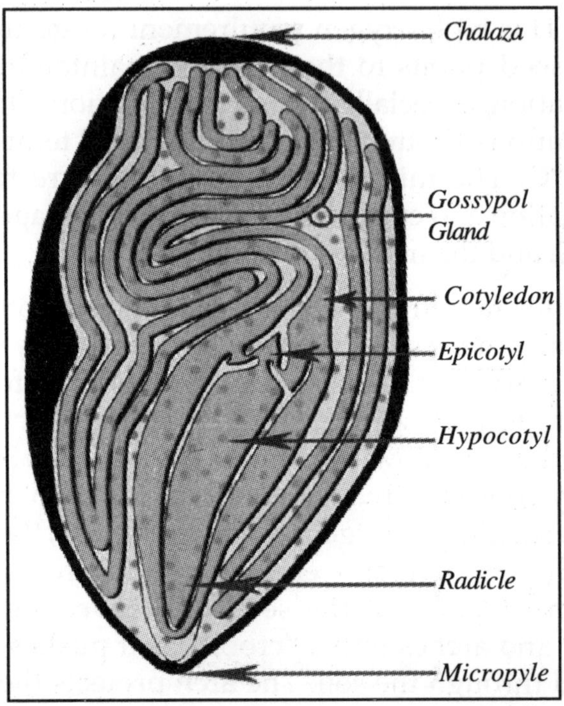

Figure 3.1. *Internal structure of seed (dormant seedling rests inside)*

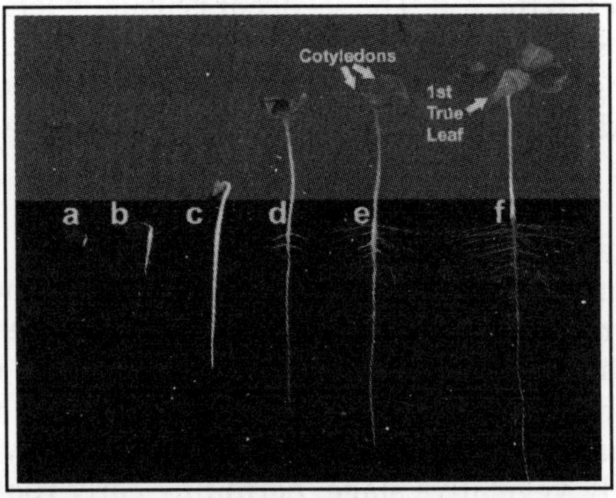

Figure 3.2. *Germination and early seedling development*

When plants begin to set bolls with a large demand for carbohydrates, root growth slows abruptly. Root length actually begins to decline as older roots die during the boll-fill period (Fig. 3.3) activity drops as roots grow older, leading to a less effective root system for uptake of water and nutrients during boll-fill is at a peak.

The primary root, or taproot, of the cotton plant is continuous with the main stem and grows without branching for several days. Secondary (or side) roots begin to develop about the time the seedling becomes erect and the cotyledons (the seed leaves) expand (Fig. 3.3). The depth to which the primary root ultimately penetrates depends on the depth and character of the various soil layers and upon soil moisture and aeration. In good soil where the above ground portion of the plant is 20.32 to 25.4 cms high, the taproot is 3 to 5 feet deep; however, a taproot 6 to 7 feet deep is not uncommon.

The concentration of the root system in the soil depends on the soil structure and permeability and on the quantity of moisture available. If the plant grows in a soil with little moisture in the upper layers but adequate moisture in deeper ones, the main root system can develop at a considerable depth; when early growth is in fairly wet soil overlaying a compacted layer, the greater part of the functional root system is usually shallow. If the taproot is destroyed, secondary and tertiary roots proliferate to form a lateral root system which does not penetrate the soil as deeply as taproot. Such plants are more subject to moisture stress later in the season. Sixty per cent of the water used by the cotton plant comes from the upper 2 feet of the soil and 75 per cent comes from the upper 3 feet.

Vegetative growth pattern

A cotton plant begins its growth with its two cotyledons (the seed leaves that form nodes opposite each other at the base of the main stem) providing the only energy (stored and from photosynthesis) for the plant to use until it grows its first true leaves (Fig. 3.3) to improve its photosynthetic "factory". The carbohydrate (sugar) production from photosynthesis is the primary energy source for growth of vegetative and reproductive organs.

Development of a healthy root system for acquiring soil nutrients is often overlooked but is equally important in supplying the plant with the necessary components for growth. Temperature determines the developmental rate of the plant, i.e. how quickly the plant adds new plant parts. Plant growth depends on a combination of (1) radiation used by healthy leaf area for carbohydrate production; (2) temperatures for rapid plant development; and (3) a healthy root system to take in the nutrients required by the basic plant structure and to feed the growing plant.

Main stem and branches : As a cotton plant begins to grow, it develops a series of nodes up the main stem. At each of the early nodes (prior to fruiting), one main-stem leaf is found. The stem will extend to the next node, with the growth between the two nodes referred to as the internode. If a vegetative branch or fruiting branch grows, it will be formed at the node. The vegetative branches typically are formed prior to the fruiting branches on one of the first four to seven nodes.

Vegetative branches have a similar morphology to the main stem and will produce their own fruiting branches. Many of the first four or five nodes will not bear any branch structure. Beginning with the fifth or sixth node, the plant begins to form fruiting branches, which bear the developing cotton fruit (square, bloom, boll). A typical cotton plant will continue to add nodes and fruiting branches for a total of 18 to 22 nodes, with 12 to 16 fruiting branches. Most of the yield is set on eight to ten fruiting branches. Under "normal" growing conditions, a new mainstem node will be formed on the plant every three days. A cotton plant can set its boll load in as little as three weeks and still have good to excellent yield potential.

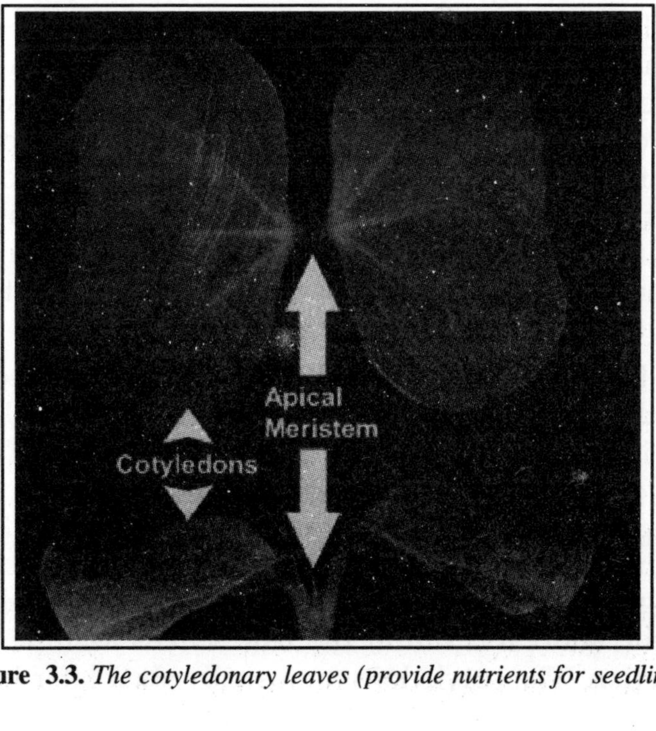

Figure 3.3. *The cotyledonary leaves (provide nutrients for seedlings)*

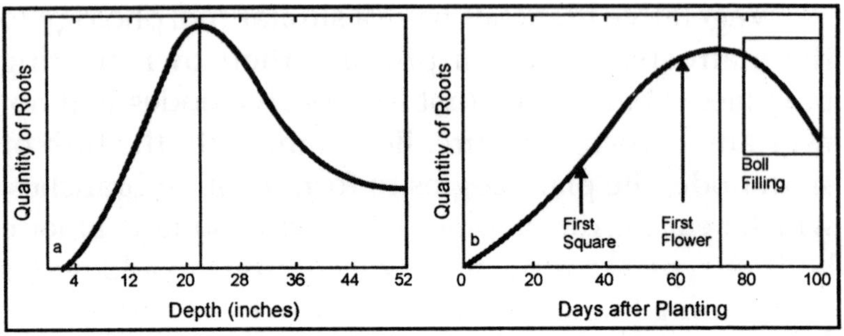

Figure 3.4. (a) Comparison of root quantity with soil depth for a mature cotton plant. New roots are constantly produced in areas of the soil profile that have water and nutrients. **(b)** Comparison of root quantity with cotton growth stage. Roots begin to decline after flowering

Leaf

The arrangement of leaves on a stem is known as phyllotaxy. Cotton leaves are arranged in a spiral with each leaf being 3/8 of a turn above the last. A stem may have the leaves arranged in a right or left handed spiral and occasionally you will see one that reverses direction. The first true leaf of the central stem is usually not lobed. As the plant matures, the leaves formed at higher positions are more and more deeply lobed until the climax leaf shape is reached at about the sixth to tenth position.

Leaves of cotton are of three types. The cotyledons are the thick, kidney-shaped, seed leaves. Prophylls are the inconspicuous, first leaves on any branch. "True" leaves are broad, thin, papery, generally hairy, and range from light to dark green. After seedling emergence, the cotyledons turn green upon exposure to light. They manufacture food through photosynthesis in the short term until the true leaves assume that function. The cotyledons are usually shed later in the season. Leaves of most cotton varieties have five more or less well defined lobes. Although leaf shape is largely a varietal characteristic, climate and culture can markedly affect leaf development, particularly numbers, thickness, and size. The leaf stalk (petiole) is usually about as long as the leaf blade.

The leaf surface contains many pores (stomata) through which gases are exchanged between the plant which contributes oxygen and the surrounding atmosphere that contributes carbon dioxide. Most, but not all, of the stomata are on the underside of the leaf.

To provide the carbohydrate energy supply for adding nodes and branches and for growing bolls, leaves function as the energy conversion factory. Photosynthesis converts light energy to chemical energy that is stored as sugars in the plant. All plant metabolic reactions are dependent on this energy source. An efficient plant will develop a canopy quickly to capture nearly all of the solar radiation available. Plants that have established a full canopy by the bloom period will be able to keep more bolls and fill them more effectively.

Glands

The cotton plant contains two types of glands, internal and external. The internal glands are visible as black dots generally distributed over the entire above ground portion of the plant. These glands contain gossypol and play a role in insect suppression. Some varieties without the glands (termed "glandless") are available. The external glands called "nectaries" are found underneath the leaf and inside as well as outside the bracts of the flower. Under the leaf, they are found on the main midrib and occasionally on one or both of the largest midribs. These glands produce a sweetish fluid that plays a role in attracting insects. Some varieties without these glands (termed "nectariless") are available.

Nodes

The point on the stem from which the leaf originates is called a node. The points where the two seed leaves or cotyledons were attached are opposite and are the first node. The point at which the true leaf occurs becomes node second and successively on up the main stem.

The true leaves develop in a regular spiral arrangement on the main stem, either to the left or to the right. The distance around the stem from one true leaf to the next above or below is normally three eighths of a complete turn.

As a cotton plant adds new nodes, it is also adding new leaves to the photosynthetic factory. The leaf area on a cotton plant is very small prior to first square. When the plant begins to square, two leaves are added at each node (one mainstem leaf and one on the fruiting branch), which results in a large increase in leaf area. Since leaves reach their maximum photosynthetic production between 16 and 21 days of age, a cotton plant is able to enlarge its photosynthetic production by adding new leaves throughout the bloom period.

The cotton plant has two types of branches, vegetative (monopodial) and fruiting (sympodial)

(Fig. 3.5). The main stem is a vegetative branch. Usually, vegetative branches of the main stem occur in a definite zone near the base of the plant, and the fruiting branches occur farther up the stem. Vegetative branches commonly range in number from zero to four. Generally, the more vegetative branches a plant has, the later in maturity it is. Once fruiting branches develop, it is rare for more vegetative branches to be formed. The number of nodes from the base of the main stalk to the first fruiting branch varies considerably among cotton varieties and is affected by plant spacing, nitrogen fertility, moisture, abortion of the terminal bud, and/or lodging of the plant. In most upland cottons, the first fruiting branch develops no lower than the fifth or sixth node above the cotyledons. New fruiting branches are produced by the terminal bud of the main stem at approximately 3-day intervals during the plant's period of active growth.

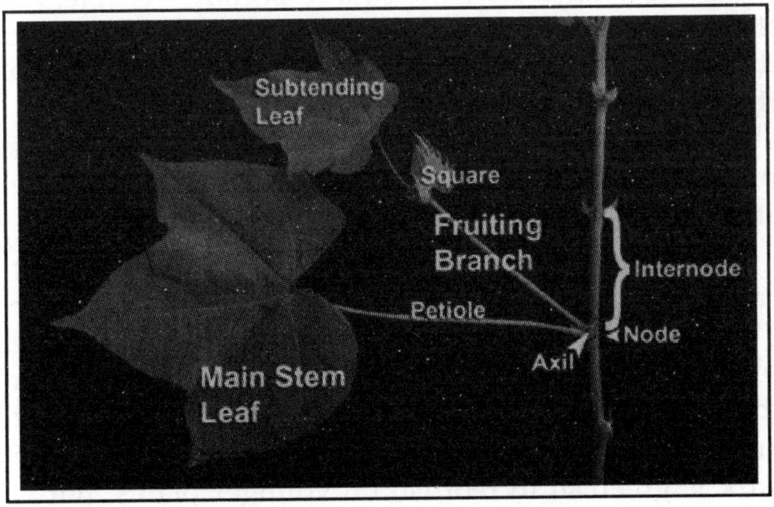

Figure 3.5. *Growth of a fruiting branch from the main stem.*

Each vegetative branch of the cotton plant has an almost identical growth pattern and develops continuously throughout the growing season from an apical bud at the top of the stem (Fig. 3.6). Some varieties tend more toward lateral growth of the vegetative branches than do others. When growing laterally, vegetative branches are likely to bend over with the weight of maturing bolls and assume an almost horizontal growth. Vegetative branches are straighter and slower growing than fruiting branches. They must rebranch before they can bear flowers and bolls.

Fruiting branches are smaller in diameter, faster growing, and assume a more nearly horizontal attitude than the vegetative branches. The first part of the fruiting branch to become visible is the floral bud (square). As the branch develops, the bud is carried away from the main stem by lengthening of its internode. A leaf develops beside the square, but remains very small until 4 to 7 days after the square becomes visible. As the leaf enlarges and unfolds, the first axillary bud at its base grows out at an angle with its floral bud to form the next internode and flower of the fruiting branch. The resulting development of fruiting branches produces a zigzag appearance in contrast with the straighter vegetative branches.

A mature plant will normally be 2 to 5 feet tall, but can range from 15.2 cms to over 7 feet. It will be columnar ("cluster" or "semicluster"), pyramidal (Christmas-tree like), or more commonly rounded in shape. Wood in the stem is relatively soft. The bark is moderately thick and tough. The outer layer is more or less corky. The bark is greenish to reddish on younger plants, but is yellowish brown on older plants.

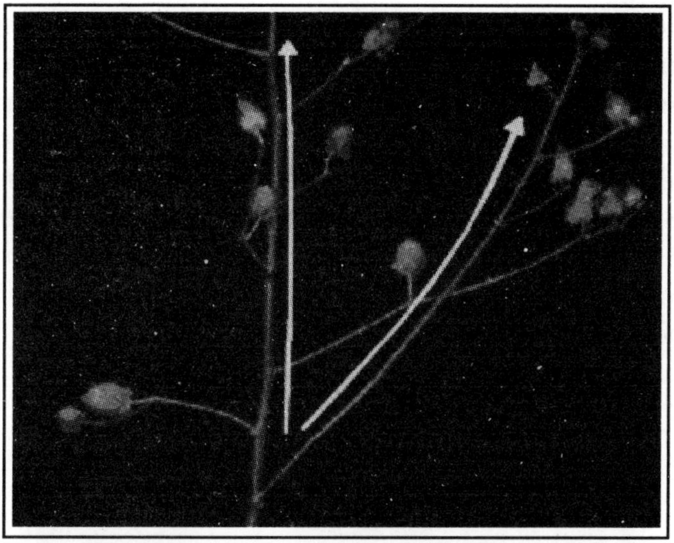

Figure 3.6. *A cotton plant with leaves removed shows the straight growth habit of the main stem and the vegetative branch.*

Reproductive growth pattern

Fruiting branches : Fruiting branches typically form above the fifth or sixth node on a cotton plant and bear the developing fruit. The developing branch terminates in a square, but a second square and leaf develop adjacent to the first leaf and extend a new internode of the branch away from the first fruiting position. The fruiting branch grows by repeating this process to produce several squares, leaves and internodes in a zigzag pattern, as illustrated in Fig.3.7.

Fruiting structures: Squares are the flower buds that first appear on the plant when reproductive growth begins. The flower bud is enclosed by three bracts. Squares grow for about three weeks before a bloom appears. During this time numerous events are occurring within the flower bud structure. Two weeks before the square is visible on the plant, the floral stimulus is initiated, and just prior to the pin-head square, the number of ovules (potential seeds) is established (Fig. 3.8). 'The male and female parts of the flower undergo their development within the square and the fiber cells begin to differentiate just prior to bloom. The actual number of seeds within a boll is dependent on effective fertilization of the bloom.

Figure 3.7. *A fruiting branch with leaves removed shows its zig-zag growth habit.*

The first square can usually be recognized in about 35 to 40 days after emergence of seedlings. Twenty to 25 days are required for it to develop into an open flower. Thus, cotton planted in mid to late April will usually produce the first flower sometime in late June. Flowers appear on successive positions of the same fruiting branch at intervals of about 6 days. The number of squares produced is directly related to the number of fruiting branches and their fruiting positions. Peak blooming usually occurs about a month after first bloom, levels off for 2 to 3 weeks, and then declines rapidly.

Flower

First encountered in a bud are a whorl of large, conspicuous, green, leafy bracts collectively termed the "involucre". The three bracts, set close together, form an upright triangular structure with the remainder of the floral bud enclosed in the center. Immediately inside the bracts is the true calyx, consisting of five short, equally lobed, green sepals fused together to form a tight cup around the broad, lower end of the bud. Inside the calyx are five petals of the corolla (Fig. 9). These are separated except at their basal point of attachment, but are wrapped tightly around each other in the bud. Inside the corolla is the staminal column, bearing about 10 double rows of filaments with two-lobed anthers.

The innermost structure is the pistil. The enlarged conical part at its base, called the ovary, consists of three to six carpels (locules). Typically, the ovary contains 8 to 12 ovules (undeveloped seed). These are initiated in two parallel vertical rows along the placenta, the central column where the carpels are joined together. In the cotton flower, the long tip of the pistil (the stigma) is normally composed of fused lobes; the number of lobes corresponds to the number of locks. The style is the column which connects the stigma to the ovary. Flowering begins low on the plant near the main stem. It proceeds spirally upward and outward on the plant.

The open flower in upland cotton is large and showy. Petals are a creamy white to light yellow in colour with no petal spot (Fig. 10). The petals open on day one, closing at the end of that day never to reopen. Toward the end of day one, the petals assume a pinkish colour; on the second day, (Fig. 11) they are reddish in colour; usually on the third day, they wither and soon fall off the tiny developing boll. Pollen of cotton is large, cream or yellow in colour, round with spines, sticky, and relatively heavy. There are about 45,000 pollen grains per flower, each grain remains viable for about 12 hours. Cross-pollination ranges from near zero to almost 70 per cent. It is accomplished by bumblebees, honey bees, solitary bees, and likely other insects, but not by wind.

The cells of the flower petals expand rapidly during the 24 hours preceding bloom, and the flower opens during the early to midmorning hours. During this time the male and female flower parts expand rapidly. The flower opens when the cotton reproductive system has reached maturity. The embryo sac is fully developed and ready for fertilization by the pollen from a mature anther at bloom. Shortly after the flower opens, the anthers split open and discharge pollen which adheres to the sticky surface of the stigma. Pollination usually occurs without the aid of insects because of the closeness of the stigma and anthers. Very little pollen is disseminated by air movement. If the ovules are not fertilized properly, the plant sheds the young boll. The flower petals turn pink on the second day and later dry up.

Boll development : It begins with fertilization of ovules and alters the status and growth of the plant. Rather than sending most of its resources to vegetative growth, the plant must now support continued vegetative growth as well as growing bolls. The growth of a boll can be divided into three phases : enlargement, filling and maturation. The enlargement phase determines the final volume of the boll and seeds, while the fibers are elongating during the first 20 days after bloom. The filling period is the next 20 days after bloom, when much of the dry matter is added to the boll through secondary wall formation of the fiber; oil and proteins are added to the seed during this time as well. The maturation period allows both the fiber and the seed to mature physiologically and dry down prior to boll opening.The cotton boll is usually light green in colour, relatively smooth, and has few oil glands. When ripe, it splits along lines where the carpels meet. The moist fiber dries and fluffs into lint in 3 to 4 days. The dried carpel wall is called the "bur". Dried bolls are of three types. The burs of "open" bolls open widely; and with prolonged exposure, locks of seed cotton tend to fall to the ground. "Storm resistant" bolls hold the seed cotton more firmly and lose considerably less to the ground.

The boll develops rapidly after fertilization and reaches full size after about 21 days. An additional 20 to 50 days elapse before the boll is mature and ready to open, depending primarily on temperature. Bolls from flowers opening during July mature much more rapidly than bolls from flowers opening during late August.

The size of the boll varies with variety, soil fertility, soil moisture, and cultural practices. When bolls mature, the carpel walls dry and split along the lines where the carpels meet. Under favourable conditions, the bolls open rapidly, and the locks of cotton dry and fluff. The number of locks corresponds to the number of carpels and varies from three to five. The number of seeds per lock is about nine.

Development of Seed

The fertilized ovule (developing seed) enlarges rapidly, reaching its maximum volume about 3 weeks after blooming. About 2 weeks prior to the opening of the boll, the seed is completely filled by the embryo, which has already become differentiated into sprout and seedling leaves. However, maximum weight is attained only a few days before the boll opens. Most of the oil and protein accumulates in the embryo during the second half of the boll maturation period.

Each cotton fiber originates from an extension of a single cell on the outer layer of the seed coat. These cells begin to elongate about the time the flower opens. They continue to elongate as thin-walled tubular structures until maximum length is reached, usually about 18 to 20 days after blooming. Fiber length is shortened considerably by a shortage of water if it occurs during the period of fiber elongation. After the fibers reach their maximum length, the fiber walls are thickened by the deposition of consecutive layers of cellulose on their inner surfaces in a spiral fashion.

Secondary thickening continues until just before the boll opens. Secondary-wall structure never completely fills the cell, as some space is always left in the center. The degree of thickening and the angle of the spirals affect fiber maturity and strength. They are dependent upon variety as well as environmental conditions which affect growth and opening of bolls. Droughty conditions during the wall thickening process usually increase fiber strength slightly.

Until the boll opens, the cotton fiber is a living, cylindrical cell. Moisture is lost from the fibers when the boll opens, causing the fibers to collapse and die. As the fibers collapse, they assume a flattened, ribbon-like from with few to many twists which help cause the locks of seed cotton to fluff.

Shedding of Fruit

The cotton plant has tremendous fruiting potential (Fig. 3.9). It produces many more flowers than it can possibly mature. Only about 40 per cent of the fruiting forms produced under ideal conditions eventually mature. This great reserve of fruiting forms effectively serves to minimize loss in yield as a result of conditions temporarily unfavourable for fruiting or from infestations of insects.

Fruit shedding may occur as a result of insect injury or as a result of conditions attributed to the physiology of the plant. This is called natural shedding. Almost all fruiting forms that shed naturally are either very young bolls or very small squares. Young bolls are more readily shed than small squares. Most squares over one-third grown normally do not shed unless injured. Shedding does not occur in the flower stage and only seldom in bolls more than 10 days old.

Factors which adversely affect growth such as drought, flooding, prolonged cloudy weather, unbalanced fertilizer, and root pruning by deep cultivation are known to accelerate shedding temporarily. When conditions become more favourable for growth, the plants compensate by retaining a higher percentage of their later blooms.

Usually, vegetative growth and flowering gradually decrease and practically cease after the plants have become heavily loaded with bolls later in the summer. The rather abrupt reduction in growth and flowering is referred to as cut-out. Apparently, the developing bolls have first claim to the plant's food supply produced by photosynthesis and the nutrients absorbed by the root system (Fig. 3.10). Growth is slowed and finally stopped and the plant matures the bolls that are already set. With the approach of cut-out, shedding becomes very high.

In addition to the young bolls, many of the very small squares are shed, especially if cut-out is abrupt. Many of the leaves on the lower portion of the plant drop during this stage. The cessation of vegetative growth and the loss of leaves on the lower portion of the plant stimulate the rapid opening of bolls and reduce losses from boll rot. Thus, the occurrence of cut-out is desirable unless it occurs so early that yield is reduced.

The degree and time of cut-out are influenced by variety as well as environmental conditions such as insect control, weather, and soil fertility. High rates of nitrogen, particularly in combination with high rainfall during late summer and early fall, are very conducive to continued vegetative growth.

Order and Rate of Development

The cotton plant develops according to a time schedule that is predictable within limits. Most developmental phases are consistent. Other phases require relatively longer versus shorter periods according to local climate, variety, and method of culture. The schedule can differ considerably between different cotton growing areas. In general, the schedule is as described below.

Under highly favourable conditions, a cotton plant will emerge as early as 4 days after planting. Under very poor conditions, it can require upto 3 weeks. Normally, emergence requires 5 to 10 with probably 7 days being a reasonable average. If an acceptable stand has not been attained in 10 days, serious consideration should be given to replanting. Usually, about 35 to 40 days after seedling emergence, the first square will appear. The development of the square into an open bloom commonly requires 23 to 25 additional days. This period of development is more consistent than the period between emergence and appearance of the first square. The interval between blooms is about 3 days for a given node on one fruiting branch and on the same node of the next higher fruiting branch.

Development from open bloom to open boll requires between 45 to 55 days for early and midseason blooms. Late-season blooms require 55 to 70 days. Variation is due primarily to temperature. Most rapid development occurs in the warmer part of the season; development can be exceptionally slow with low fall temperatures.

Squares can be shed at almost any age, but most fruiting forms that are shed do so prior to blooming. Most early fruit shed is due to insect damage. Time of shedding is affected by drought, extremes in temperature, cloudy weather, insect injury, and disease. Shedding in response to injury may occur from 3 to 10 days after the injury actually takes place. Boll shedding usually occurs within 7 days after blooming. This interval also ranges from 3 to 10 days. Bolls older than 10 days are seldom shed except when the plant is subjected to extreme shock such as harvest-aid chemical treatment, severe water stress, boll weevil or bollworm attack, or frost.

Growth pattern adjustment

When a cotton plant can no longer supply the needs of both vegetative and reproductive growth, it adjusts its growth pattern. Vegetative growth gradually slows and then stops. However, some short-term deficiencies are relieved by reducing the number of fruit on the plant. Fruit structures with the lowest priority and least investment are aborted i.e : squares and small bolls. This reduces the immediate demand for carbohydrates and nutrients. Squares require very little carbohydrate from the plant, but they represent future demand when they become bolls. Bolls are a much higher priority than squares, because the plant already has a large investment in medium to large bolls and is biologically programmed to survive by producing the seed sound in the boll.

The fruit shed phenomenon, particularly a physiological fruit shed and not fruit loss due to insect feeding, is typically found after bloom and after cloudy days. Reduced radiation on cloudy days reduces the carbohydrate supply of the plant and results in squaqres shed from the plant. If shedding squares does not reduce the nutritional shortfall, then small bolls are shed. Cotton plants never shed blooms and rarely shed bolls more than seven days old.

Fruit shed is of particular concern when it occurs early in the fruiting cycle and potentially delays the maturity of the crop. Measuring fruit shed and managing to avoid problems with fruit shed are important strategies and one of the primary purposes of plant mapping.

Heat unit requirements: Cotton develops at a pace determined by the temperatures that the plant encounters. If temperatures are between 18.3°C and 29.4°C degrees, a cotton plant develops at a very predictable rate. In other words, above a base temperature of 15.56°C degrees, a cotton plant grows. Below that threshold, little or no development occurs. The calculation of heat units is very straightforward for the conditions encountered in the region of the Cotton Belt. The calculation is made by averaging the maximum and minimum temperatures for a day and subtracting the base temperature to find the heat units. It is illustrated as follows :

$$\text{Heat units} = \text{Maximum temperature} + \text{minimum temperature} - 15.56°\text{C}$$

A study of heat units can be helpful in managing cotton crop's development throughout the season. Table 1 provides an outline of the heat unit requirements for a cotton crop to help you assess your crop's rate of development.

Table 1 The average number of heat units required for various growth stages of Cotton

Development stage	Heat Units (DD-60s)
Planting	0
Seedling emergence	55
Add nodes to the main stem	45 to 65/node
First square	500
First bloom	850
Cutout (assuming 20 mainstem nodes and 10 to 12 effective fruiting branches)	1,300 to 1,450
First open boll	1700

Plant mapping

By quantifying several growth parameters of the cotton plant, growers and consultants can identify potential problems or opportunities for managing their cotton. This management approach requires the time and effort to collect and interpret these simple plant maps. Early approaches to plant mapping were a very intensive effort that required every fruiting site on the plant to be recorded as either a square, bloom, boll or aborted position. This was very similar to the plant maps used by cotton researchers. Further, studies have shown that a few easily collected numbers can quickly provide much information for managing cotton.

Measurements

Sampling a cotton field usually calls for going to four areas of a field and measuring five plants in each area. Three sampling periods are important : prebloom, bloom and postbloom. The measurements taken before bloom are important in measuring the early growth pattern and avoiding early problems that slow development of mainstem nodes and leaf area. During the bloom period, the primary concerns are retention of fruit on the plant and how quickly the plant is moving to "cutout". After cutout and while bolls are opening, correct timing of harvest aids can be determined by some simple plant mapping techniques.

Node of first fruiting branch: To assess the shift from vegetative growth to the beginning of fruit development, count the number of nodes to the first fruiting branch. During early squaring, this is the first node that has a square. Count the nodes by finding the cotyledons. Begin counting the first node above the cotyledons that was formed by a true leaf.

The first fruiting branch will typically form on the fifth or sixth node on cotton. Forming lower than the fifth node is unusual and may lead to the cotton plant cutting out too quickly, since it may not have enough leaf area to support the fruit forming on the plant. If the first fruiting branch is delayed to the seventh or eighth node, then the beginning of the fruiting cycle is delayed, which may require a shorter fruiting period if the late fruit on the plant is unable to mature in a short season. Avoiding insect damage or other stresses are good management strategies to help the cotton plant begin fruiting on an optimal node.

Height to node ratio : To assess the vigour of the cotton crop, measure both the height and node number of several cotton plants and calculate the ratio of height to nodes by dividing (plant height divided by the number of mainstem nodes). This ratio indicates the amount of stress that a cotton plant has encountered, because the development of nodes is not influenced by stress (before boll set) while plant height is greatly influenced by various stresses. In other words, the number of nodes is the age of the crop, and the height is an indicator of stress encountered. The target value for this ratio will vary according to variety and time of the season, but the following values can serve as general guidelines. Plant height is closely related to canopy size and photosynthetic capacity of the plant. Using these guidelines of height to node ratio should lead to canopy closure by mid-bloom, to optimize available sunlight.

Fruit retention : Throughout the reproductive growth phase, the cotton plant is constantly adding squares to the plant and then aborting squares or young bolls to balance out the demand of the growing boll load. By counting the number of fruiting branches above the uppermost first position white flower (with squares) & the number of squares at the first position sites, the per cent square retention can be calculated. Similarly, by counting the number of nodes below the uppermost first position white flower and number of bolls at the first position sites, the per cent boll retention can be calculated.

Per cent retention for squares prior to bloom should remain very high, since squares only need small quantities of carbohydrate for their survival. Physiological square shed does not occur until the demand for carbohydrate is peaked by growing bolls. Square shed prior to first bloom may be caused by insect feeding, particularly by plant bugs or boll weevils. If square loss before bloom is noticed, a look at insect pressure is needed.

Early square shed can also delay maturity, which may require a closer watch on the maturity of the crop as it approaches cutout. Boll retention should begin near the level of square retention and show a gradual decline throughout the bloom period as the plant reaches its capacity for supplying bolls with carbohydrates.

Young bolls are sensitive to physiological shed when carbohydrate supplies are limited due to cloudy weather, high temperature, water stress, leaf damage or a heavy boll load. During this stage, management options should help the plant keep the supply of carbohydrate large, to permit the boll load to grow and supply all the carbohydrate need for boll growth. Avoid problems that can reduce carbohydrate supply by timely irrigation, good fertility and avoiding stresses that would reduce the canopy size.

Nodes above white flower (NAWF) : A very accurate and sensitive measure of the cutout of cotton is the number of nodes from the uppermost first position white flower to the termination node of the main stem. As the fruiting cycle of cotton reaches its peak, the plant cannot meet the demands for both unlimited vegetative and reproductive growth, so the plant slows & eventually stops adding mainstem nodes. The squares that were formed with each node eventually form blooms that are progressively closer to the terminal.

Nodes above cracked boll (NACB) : As the crop approaches maturity and time for harvest aid application, the opening of bolls up the cotton plant mimics the pattern of white flowers blooming up the plant prior to cutout. The NACB is determined by counting the number of nodes from the uppermost cracked open boll to the uppermost harvestable boll. Counting the number of unopened and opened first position bolls and calculating per cent open bolls is recommended. More than 50 to 60 per cent of bolls should be open before applying harvest aids to avoid yield and fiber quality losses.

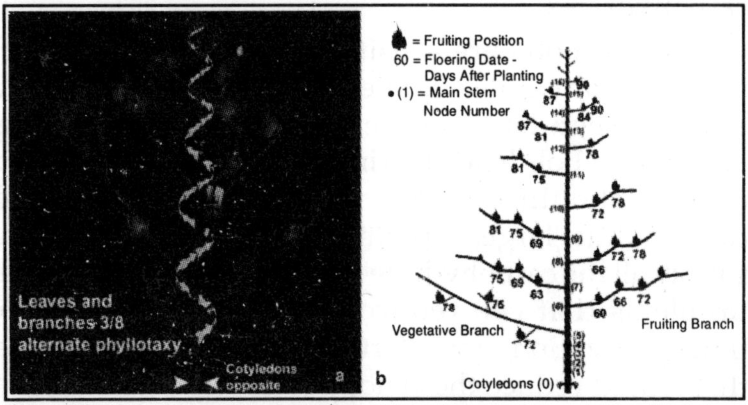

Figure 3.9. *A diagram of the general timing of flower emergence from buds on the fruiting branches by fruiting position.*

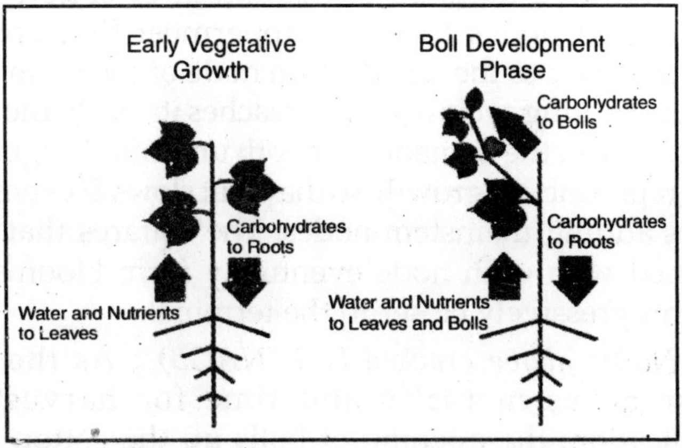

Figure 3.10. *During early vegetative growth, most of the carbohydrates produced by the leaves are sent to the root system. Later in the season, how-ever, most of the carbohydrates are sent to the developing bolls, and the root system and shoot growth rate decline.*

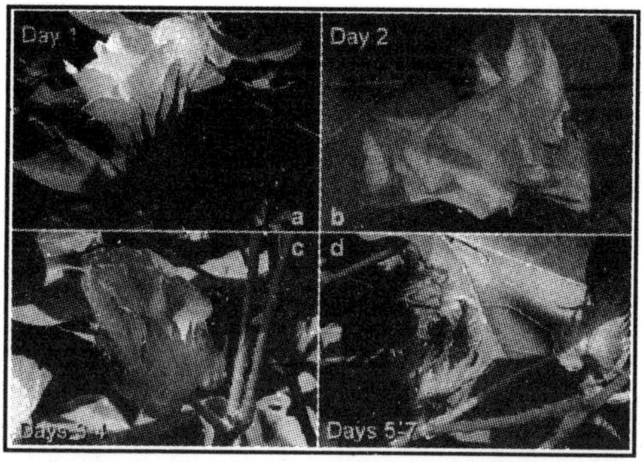

Figure 3.11. *Development of a cotton bloom. A white flower emerges on day 1 (a), then gradually darkens and takes on a red color during days 2, 3 and 4 after emergence (b & c). The bloom eventually dries up and either falls off*

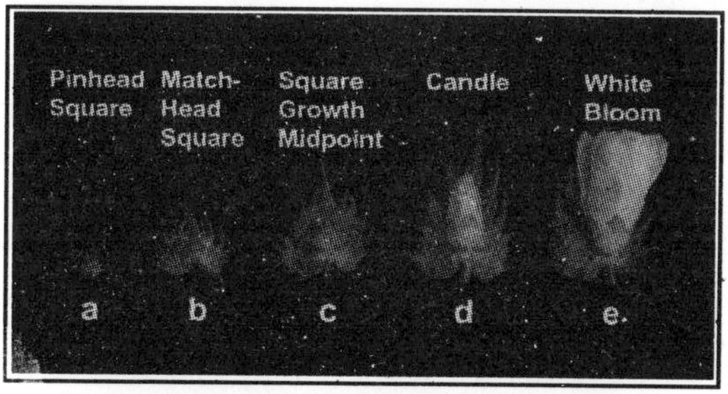

Figure 3.12. *Parts of a mature cotton flower.*

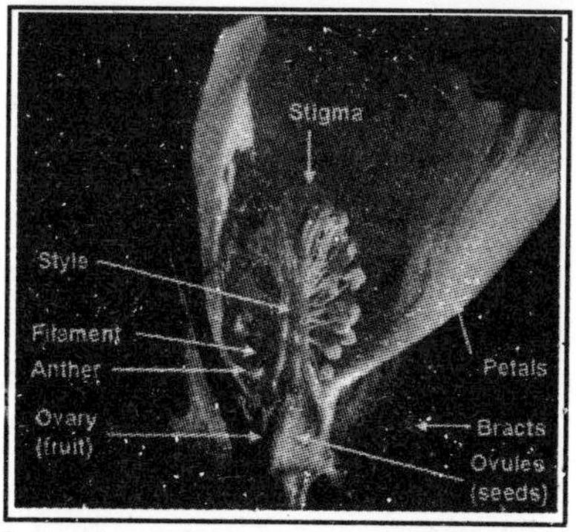

Figure 3.13. *Development of the bud from match head square (a) to flower (e). Two bracts have been removed from each square, candle and bloom to show this development.*

□□□

4

Physiology of Crop Maturity Defoliation and Harvest-Aids

A basic knowledge of crop development and maturity along with an understanding of the physiology of harvest-aids is necessary in making decisions concerning the effective application of these materials. Successful preparation of a cotton crop is desirable for harvest factors in the complexities of crop, leaf senescence, boll maturation and the diversity of harvest-aid functions.

Crop Condition and Maturity

Cotton that has depleted residual soil nitrogen and water, has ceased flowering, and has a good boll load, generally defoliates readily. A crop in this condition actually begins to shed lower leaves naturally. If the application of harvest-aids is delayed significantly beyond the time of crop maturity, or if there is an abundant supply of soil nitrogen and water, re-growth or second growth often occurs. Re-growth or second growth consist of juvenile leaves emerging late in the season and such growth is much more difficult to remove from the plant than older, more mature leaves.

Additionally, vigorously growing plants with limited boll loads are more difficult to defoliate than mature, heavily fruited plants. Re-growth is especially challenging if late season rains follow early or midsummer drought.

The development of the cotton plant and its maturity are determined primarily by temperature, water, nutrients, and sunlight. The effect of temperature on the cotton plant is quantified by the calculation of heat units or degree-days above 15.55°C (DD60's). A base of 15.55°C is used since below this temperature cotton growth basically ceases.

Daily DD60 = (Daily Maximum Temp (°C)) - 15.5°C

The accumulated DD60's for a given time period can be calculated by summing the daily DD60's over that time. Temperatures above 32.22°C should be entered into the formula as 32.22°C, since growth is not expected to increase significantly above this temperature. The table below lists the average DD60's and the range in number of days required to reach certain growth stages.

From Planting to	DD60's	Days
Emergence	50	4-14
Pinhead square	550	35-45
First bloom	940	55-70
Peak bloom	1700	85-95
First open boll	2150	115-120
Harvest	2500-2700	140-160

Boll Development and Maturity

For a boll to fully mature, 900 to 950 DD60's are required after pollination. Minimum boll maturity, or the earliest time that harvest-aids may be applied without affecting yield or quality, can be achieved with 750 DD60's. Thus, under optimal conditions during the early part of the fruiting cycle, a boll matures in 45 to 50 days after flowering. Blooms occurring in late August or early September often require 65 to 70 days to develop into a mature boll. This lengthened time requirement for late-season bolls is due to relatively lower temperatures in the fall versus late summer, subsequently resulting in fewer DD60's accumulated per day.

The cotton plant flowers and set bolls for 4 to 8 weeks. In general, bolls set during the first 5 to 6 weeks of the fruiting cycle are the largest and most mature. Under optimal growth conditions, 95 percent of the crop can be set in 4 weeks. Typically, the smaller, less mature bolls set later in the season make up only a small portion of the total crop. However, under conditions associated with drought or severe heat these later bolls can make up a significant portion of the crop when earlier bolls are aborted.

The development of the boll can be divided into three phases. The first stage is the fiber elongation phase, which lasts about 3 weeks. Each fiber is a single cell originating from the outer surface of the developing seed. During the elongation phase, the fiber increases in length and resembles a straw in structure. The second phase is fiber filling, which lasts about 3 weeks. Through the fiber filling phase, cellulose is deposited inside the hollow elongated fiber cell thereby increasing its thickness. This filling phase is similar to the growth rings on a tree; however, unlike a tree, the cellulose is deposited inside the old growth rather than outside. Fiber filling adds weight and thickness to the developing fiber. The thickness of the fiber is ultimately measured and reported as micronaire. If this process is interrupted by the application of harvest-aids or unfavorable environmental conditions, low micronaire results and the fibers weigh less than if they were allowed to develop fully.

After the filling phase, the boll maturation phase begins. During this phase, boll dehiscence occurs. The capsule walls of the boll dry causing the cells adjacent to the dorsal suture to shrink unevenly. This shrinking then causes the suture between the carpel walls to split, the boll opens and fibers dry. A sharp knife can be used to determine boll maturity. Mature bolls are difficult to cut in cross section with a knife without stringing the fiber. Boll maturation is also correlated with seed maturation. The seed coats of mature seeds are tan to brown as opposed to the white and pale color of immature seeds. Inside the mature seeds, tiny leaves are easily identified. Immature seeds contain gelatinous material instead of leaves. Green bolls that crack when firmly squeezed should open within 7 days during warm, dry weather.

Function and Rate of Application of Harvest-Aids

Functions

Harvest-aids perform several functions in the production of .cotton. As defoliants, they aid in the removal of both mature and juvenile leaves, eliminating sources of trash as well as stains and facilitating machine harvesting. Certain products can also be used to accelerate boll opening, facilitating once-over harvest and minimizing yield and quality losses associated with delayed or extended picking time. Some harvest-aids suppress re-growth, which sometimes occurs after all mature leaves have been removed. Desiccants can be used to kill dry leaves remaining on the plant after defoliation and to dry down weeds prior to harvest.

Timing of Application

Timely application of harvest is necessary and is dependent upon crop maturity. Applying these chemicals either too early or too late can affect lint yield and quality. There are several methods that can be utilized to determine the appropriate time for harvest-aid application including; 1) percent open boll counts, 2) nodes above cracked boll (NACB), and 3) boll maturity check with a sharp knife. A combination of these methods provides the best estimate of crop readiness.

After application of harvest-aids, defoliation usually occurs within 7 to 10 days when plant conditions are favorable and temperatures are above 65 degrees F. This time frame may be reduced to 4 or 5 days under extremely favorable conditions but may also be delayed to 15 days under cooler, wetter conditions. Rapid leaf drop may speed the opening of mature bolls but delay or prevent the opening of immature bolls, since the development of immature bolls ceases once leaf function is disrupted.

A minimum of 75 percent open bolls is generally a safe time to apply harvest-aids, assuming the youngest bolls on the plant to be harvested are mature. Plants uniformly fruited from top to bottom can be treated safely with as few as 60 percent open bolls. Since the bolls set on the cotton plant open in the same order in which they are set, NACB counts can be used to determine crop maturity and timing of harvest-aids. NACB is determined in the following fashion. First, locate the uppermost first position boll that is cracked or open. Then locate the uppermost unopened boll that is to be harvested. Count the number of nodes separating the two starting at the cracked boll, counting it as 0 (Figure 4.1). A cotton crop can safely be treated with harvest-aids if the NACB is equal to 4. Under optimal conditions a boll should open on a successive node every 3 days. In addition, studies conducted have shown that there is a linear relationship between percent open bolls and NACB.

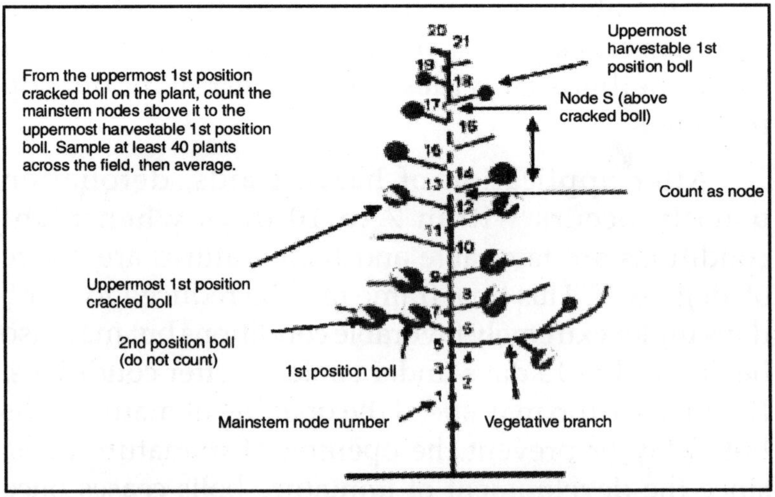

Figure 4.1. *Determining nodes above cracked boll.*

Sharp knife technique can also be used to assess the maturity of unopened bolls. As previously described, if the seeds in an unopened boll have a tan to brown seed coat and contain developed cotyledon leaves inside (Fig. 4.2), it is mature and can safely be treated with harvest-aids without loss of lint weight or fiber quality.

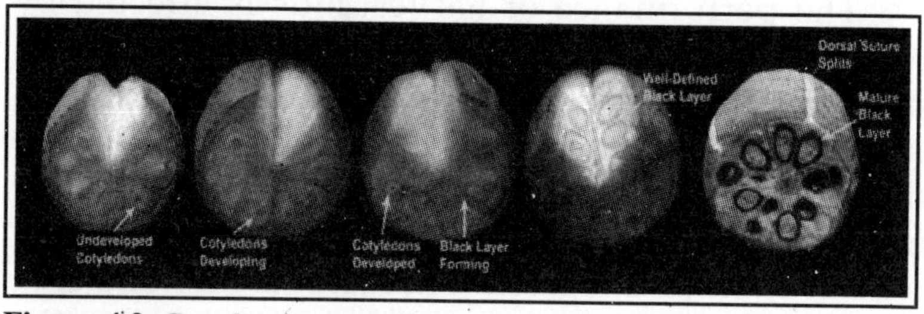

Figure 4.2. *Development of the cotton boll. Undeveloped cotyledons have a jelly-like consistency. As the cotyledons develop, they become firm and leaf-like, and a black layer that will become the seed coat begins to form around the seed.*

Timing and Profitability

The effects of harvest-aid application timing and subsequent harvest on lint yield, quality and profitability have been recorded. The studies indicated that maximum lint quality was attained when harvest-aids were applied between 40 and 60 percent open boll. However, maximum lint yields and profitability are not achieved until the crop reaches 76 to 90 percent open boll. Under the current cotton marketing system, the penalty for reduced yields is much greater than the penalty for reduced quality. Thus, managing harvest-aid application timing to maximize fiber quality does not appear to be profitable if yield is sacrificed in the process. The profitability was shown to decrease dramatically if harvest-aid application was delayed until after the crop reached 100 percent open boll. These losses in profitability were attributed primarily to crop weathering due to rainfall, which led to below base color grades and yield reductions.

Rate and Efficacy

The performance of harvest-aids is affected by temperature, plant condition, spray coverage, and product rate. Temperature is the primary force in determining harvest-aid rate. The goal is to maximize the amount of leaves that fall from the plant while minimizing the number of leaves that are desiccated or remain attached in a green condition. As the number of green leaves remaining on the plant increases, the potential for high moisture trash and leaf stain also increases. The lower end of the use rate is advisable during the early part of the defoliation season to minimize desiccation or leaf "sticking." On the other hand, cooler temperatures usually require higher rates for the desired response. Increasing rates may also be necessary if the crop is growing vigorously and there is an abundance of immature leaves. A plant that is "cut-out" has primarily mature leaves and usually requires a lower rate.

Harvest-aid application timing should not only be in concert with crop maturity, but also with picker availability. Under optimal conditions the crop could be ready for harvest within seven days after application. Ideally, the cotton should be harvested within ten days after leaf drop to avoid lint discoloration due to re-growth and weathering, and also to prevent losses due to storms. As the season progresses, the interval between application and harvest may stretch to fourteen days to allow for slower leaf drop due to cooler temperatures. Therefore, acreage treated with harvest-aids should be correlated with picker capacity. Where harvest capacity is questionable, consider beginning prior to the time that the crop can be once over harvested, and plan to second pick that acreage. This practice may improve grades and prevent losses if unfavorable weather shortens the harvest season.

Preconditioning

Preconditioning is the application of low rates of specific harvest-aids in an effort to "condition" or "prepare" the crop for defoliation. There are two situations in which this approach may be beneficial. One situation may be when plants are vigorously growing at the end of the season and the canopy consists of many green, lush leaves - a preconditioning treatment may be used one to two weeks prior to the application of the final treatment. These treatments may also be useful if there is an abundance of re-growth due to late season rainfall. Preconditioning may also be used in an attempt to partially defoliate or shed older leaves when boll rot likelihood is high. However, with this practice comes the risk of shedding too many leaves and halting boll maturation, or not shedding enough to impact the occurrence of boll rot. Generally, the risk outweighs the potential benefit.

Boll Rot and Bottom Defoliation

In theory, bottom defoliation of rank cotton could reduce boll rot when rainy weather coincides with boll opening; however, there are equipment limitations. Specialized boom shields along with near perfect timing is needed and results are inconsistent. This practice probably works the best when used in conjunction with two-stage harvesting, picking the lower open bolls prior to weathering losses and the upper bolls later. For these reasons, bottom defoliation is not recommended. If insect pests are under control, the crop is still blooming, and soft immature bolls are present, it may be best to delay application of harvest-aids and continue management practices to make a top crop to offset boll rot losses. On the other hand, if most of the healthy bolls are already open, and insect pest management has been stopped preventing development of a top crop, then harvest-aid application should proceed. Applications should also proceed if there is insufficient time for a late crop to mature.

Under normal conditions, desiccants (e.g high rates of paraquat) should not be applied with other harvest-aids. With cool weather or vigorously growing cotton, a preconditioning treatment may aid in the defoliation process. Alternatively, moderate rates of paraquat may help under these adverse conditions. When mixtures of defoliants and desiccants are applied, all bolls should be fully mature, preferably open.

Re-growth and Desiccants

Re-growth is a term given to new leaves produced by a cotton plant after stress-induced cut-out or defoliation. These are juvenile leaves that are actively growing and are often difficult to remove from the plant. Re-growth may occur in or near the plant terminal

(terminal re-growth), basal re-growth occurs along the lower portions of the plant stalk. If left remaining on the plant at harvest, these immature leaves can impart a green stain to the lint and create moisture problems in modules. Generally, terminal re-growth is more troublesome to the harvest operation than is basal re-growth. Problems with re-growth can be curtailed with timely defoliation and harvesting. In addition, harvest-aid mixtures including the active ingredient thidiazuron can also be used to reduce or prevent the occurrence of re-growth.

In certain years weather patterns are such that re-growth may be a problem despite a producer's best efforts. If this is the case, tank-mixtures containing the active ingredient thidiazuron are the best option for removing the juvenile growth. Another tactic that may be employed is the use of desiccants to dry up the leaves if they occur after regular defoliation. If desiccants are used, harvest should begin once the re-growth wilts and before it becomes brittle, this generally occurs one to two days after application. However some products require an interval of at least seven days between application and harvest. If affected foliage becomes brittle, these desiccated leaves are crushed during the harvest operation and impart fine trash into the harvested lint, thus reducing the grade.

Method of Application

Harvest-aids do not translocate throughout the plant or from one leaf to another. Therefore, thorough spray coverage of each leaf is essential in order for the abscission process to occur. Generally, hollow cone nozzles provide the best penetration and full canopy coverage, but drift can be exacerbated with the use of these nozzles in windy conditions. In addition, because of the necessity of adequate spray coverage, a second application may be necessary on rank cotton with dense foliage, since initial coverage may not be sufficient on the lower portions of these plants. If insecticide sprayers are used for defoliation, reduce the speed of the sprayer to increase the volume. Otherwise, replace the insecticide nozzles with larger tips.

Leaf Physiology and the Mode of Action of Harvest-Aids

Defoliation or leaf abscission is a natural plant process. The problem is this natural leaf drop does not occur simultaneously throughout the plant canopy, or in time to effectively facilitate mechanical harvest and at proper time. Therefore, producers must manipulate the plant to drop its leaves in a relatively short period of time to vacate the field for next crop.

While the leaf abscission process is quite complex, it can be simplified as being governed by two major hormones within the plant, auxin and ethylene. Auxin is a growth-promoting hormone that stimulates leaf growth and development. Ethylene can be classified as a senescence or ripening hormone that causes leaf drop. Leaves fall from the plant once ethylene moves from the leaf blade to the base of the petiole and stimulates the formation of an abscission layer (Figure 4.3). The amount of auxin or ethylene present in the leaves of the cotton plant is related to leaf age. This is why older leaves are more conditioned for defoliation than younger leaves. Furthermore, because of the hormone balance of younger leaves, low rates of harvest aids often have no effect, and higher rates may actually kill the leaf, leading to desiccation and leaf sticking.

Eventually, almost all the leaves on a cotton plant age so they will abscise naturally. However, producers can manipulate these hormone levels so all the leaves abscise at the same time. When harvest-aids are applied ethylene levels artificially increase so the abscission process begins.

All cotton harvest-aids can be classified into two modes of action, herbicidal and hormonal. Herbicidal harvest-aids injure the leaf, stimulating the production of ethylene. Hormonal harvest-aids increase the ethylene concentration in the leaves without causing any injury.

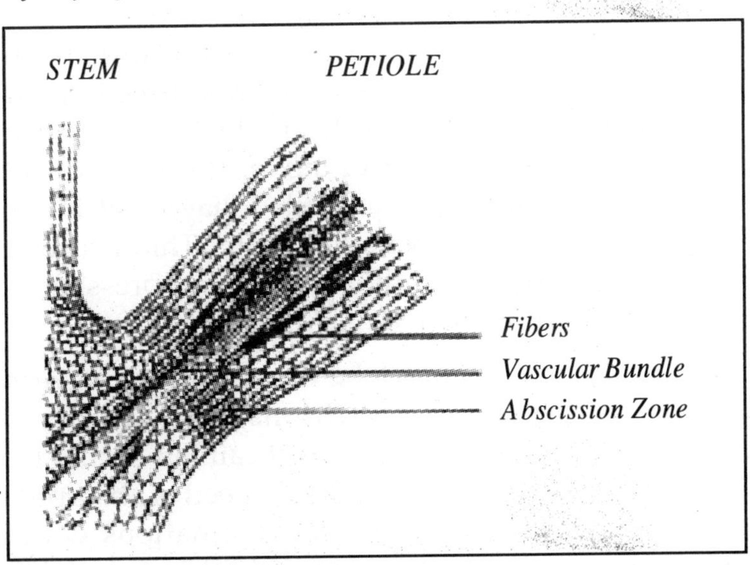

Figiure 4.3. *The abscission zone (layer) forms at the base of the petiole.*

Compounds used as harvest-aids

Like many other chemicals used in crop production, harvest-aids can be applied at various rates and in several different combinations with others. The rates and combinations are primarily a function of crop and weather conditions. Following chemicals are currently used as cotton harvest-aids :

Tribufos: It is a phosphate defoliant and has a contact-type mode of action. Once applied, tribufos injures the layer of cells immediately underneath the cuticle of the leaf. This injury then causes stress on the plant, thereby stimulating the production of ethylene. The enhanced production of ethylene ultimately leads to abscission. Tribufos is excellent material for removing mature leaves and has some limited activity on juvenile leaves. It does not function as a boll opener or inhibitor of re-growth.

Thidiazuron: Thidiazuron has a hormonal-type mode of action. This material does not directly injure the leaf but instead enhances the production of ethylene and inhibits the transport of auxin. It can effectively remove both mature and juvenile foliage, but it does not provide any boll opening activity. Thidiazuron is the best material presently available suppressing re-growth and removing juvenile foliage.

The addition of a phosphate defoliant can increase activity under cooler conditions. The activity of thidiazuron can also be reduced if rainfall occurs less than 24 hours after application. The speed of absorption of thidiazuron is increased in combinations with a phosphate defoliant or the spray additive ammonium sulfate. Of all the harvest-aid products, thidiazuron requires the longest rainfree period; in other words, its absorption into the plant is the slowest. Thidiazuron uptake is also slower in drought-stressed cotton. In this situation, higher rates or additives such as crop oil concentrate may be necessary.

Dimethipin: Dimethipin has a hormonal-type mode of action. It inhibits protein synthesis responsible for stomatal control. This loss of stomatal control causes leaves to rapidly lose water stressing the plant, ultimately leading to the production of ethylene. Crop oil concentrate can be used alone or in combination with other harvest-aids. Dimethipin is most effective in removing mature foliage from the cotton plant. It has little activity on juvenile growth and has no boll opening or re-growth control properties. However, dimethipin is a good material for the desiccation of mature morning-glories and is one of the products of choice in this situation. Data also indicates that desiccation of morning-glories with dimethipin may be enhanced with the addition of a ethephon-containing materials.

Carfentrazone: Carfentrazone is a fast-acting herbicidal-type harvest-aid. Carfentrazone inhibits one of the steps in the synthesis of chlorophyll. As a result, reactive oxygen species build up in the plant leading to the destruction of cellular membranes. This stress then causes ethylene production to increase and leaves to subsequently abscise.

Carfentrazone effectively removes both juvenile and mature foliage; however, juvenile leaves are more prone to be desiccated. It provides no boll opening activity or re-growth inhibition and is also effective in the desiccation of some weeds, particularly morning-glories.

Pyraflufen Ethyl: The mode of action, use, characteristics and performance of this harvest-aid are similar to carfentrazone.

Paraquat: Paraquat is a non-selective herbicide that is primarily used as a weed and crop desiccant; however, it can provide some defoliation and boll opening activity. It is commonly used in low rates in combination with other harvest-aids to aid in the desiccation of weeds. With excessive rates, leaf sticking and damage to green bolls may occur.

Its low rate usually avoids crop desiccation and problems with bark. A second approach is to defoliate the crop first then to follow up with a paraquat application to desiccate green weeds and foliage. Paraquat is also useful for the desiccation of re-growth as long as picking proceeds within a few days after application.

Chlorates: Sodium chlorate is distributed under many trade names. As with paraquat, this material is generally a crop desiccant. This material should not be combined with the phosphate harvest-aids or insecticides since the mixture may be explosive.

Glyphosate: The primary uses of glyphosate as a cotton harvest-aid are for re-growth control and weed management. Timing of application for glyphosate varies by cotton variety. As a re-growth inhibitor, glyphosate is only effective on non-Roundup Ready

(RR) cotton. On non-RR cotton, glyphosate can be applied once 60 percent of the harvestable bolls are open as a weed management tool and re-growth inhibitor. On RR cotton, glyphosate can be applied once 20 percent of the bolls are open as a weed management tool.

Ethephon: Ethephon is the active ingredient in many materials. Ethephon has a hormone-type mode of action. In fact, ethephon directly increases plant production of ethylene, which leads to leaf drop and boll opening. While boll opening is the primary benefit gained from the use of ethephon, this material also provides some defoliation. It has no activity on re-growth and limited activity on juvenile growth. In some situations, the addition of ethephon in tank-mixtures decreases the time required for the defoliation response to occur; in other words, ethephon combinations often provide quicker leaf drop than other harvest-aid combinations.

Ethephon enhances the opening of mature bolls. Appropriate timing of ethephon application usually coincides with sixty percent open bolls or 4 NACB. The safest time to apply ethephon is when no more than ten percent of the bolls are immature. Under certain circumstances, an earlier application may be justified such as when trying to salvage a crop ahead of a potential freeze.

Other times in which ethephon may be used include preconditioning and after regular defoliation. As a preconditioning treatment, ethephon can be applied at least four days prior to defoliation. In situations where cotton foliage is dense or the crop is lodged, ethephon may be sprayed seven to twelve days after defoliation to insure good boll coverage.

Physiology of Cotton Defoliation

It is important to understand the basic physiological processes involved in order for best crop management practices to accomplish a successful defoliation. The objectives of defoliating a cotton crop can be simply stated as :

1. to remove leaves to facilitate mechanical picking,
2. to maintain the quality of the lint, and
3. to complete defoliation with a single application of chemicals.

For this discussion, it would first be appropriate to define two terms frequently used interchangeably, defoliation and senescence.

- Defoliation, or Leaf Abscission, in cotton is usually a result of maturity, senescence, or injury. Defoliation differs from desiccation in that the leaf does not desiccate – the water conducting tissue remains alive until the leaf drops.
- Senescence is the process of plant tissue deterioration that accompanies aging and that ultimately leads to the death of an organ or organism.

Induced Defoliation

Defoliation is a natural physiological process, which usually is inadequate or not timely enough for a complete mechanical cotton harvest. Therefore, defoliation is often induced in cotton by managing the crop so that senescence, abscission layer development and leaf drop are encouraged.

Crop senescence and defoliation are encouraged most heavily by a substantial fruit load. Maturing bolls are the first priority in the plant's allocation of nutrients. As a heavily-loaded cotton plant matures, it will transfer nutrients from the leaves to the bolls. This helps to complete the maturation of the bolls, but it also makes the leaves easier to remove with a defoliant. Plant water relations, nitrogen (N) management and defoliant applications are three other crop management factors that affect crop defoliation.

Why Induce defoliation?

Pima cotton is very difficult to defoliate because of its strong indeterminate pattern of growth and its characteristic robust, late season growth tendencies, thus, considered as "worst-case scenario". Still, good defoliation is desirable in Pima cotton because of steep price discounts for lower grades of lint.

Changes Occurring during Plant Aging and Senescence

1. *The leaves lose Ribonucleic acids (RNA) and chlorophyll:* RNA are very important in protein synthesis. Chlorophyll is the pigment that makes the plants look green, and it also captures the light energy from the sun to produce chemical energy in the form of carbohydrates.
2. *Protein, carbohydrate and inorganic ion levels in the leaf drop:* The plant converts proteins and carbohydrates

in the leaves into simpler forms and then transports them along with inorganic ions out of the leaves to the bolls, which are the highest priority locations or sinks.

3. *Anthocyanins increase in the leaves* : The anthocyanins are coloured pigments that commonly occur in red, purple, and blue flowers. They are also present in various other plant parts.
4. *Hormone concentrations change* : The environment in which a plant is growing plays an important role in the hormone balance within the plant. Hormones (Table 1) affect defoliation by stimulating or inhibiting the production of enzymes

Table 1: Groups of Plant Hormones and their Effect on Defoliation

Auxins [Indole-acetic acid (IAA)]	Retard abscission
Abscissic Acid (ABA)	Promote abscission
Ethylene	Promote abscission
Gibberellins	Variable effect
Cytokinins	Variable effect

Auxin-Gradient Theory of Abscission

The auxin-gradient theory of abscission holds that the gradient, or relative concentrations of auxins in the plant, is more important for defoliation than the absolute amount of auxins present in the plant. The auxin, indole acetic acid (IAA) inhibits leaf abscission when it is concentrated on the petiole side of the abscission zone. This same auxin stimulates abscission when it is concentrated on the stem side of the abscission zone. The gradient can be affected by ethylene and abscissic acid.

Ethylene and Abscissic Acid

Ethylene and abscissic acid (ABA) are the plant hormones that commonly promote leaf abscission. It would be worthwhile to look at their effects in more detail.

Ethylene

Indirect effects

- Stimulates the activity of IAA-oxidase system, reducing the auxin concentration and promoting abscission.
- Decreases and inhibits auxin transport, promoting abscission.

Direct effects

- Stimulates the synthesis of pectinases and cellulases and increases the secretion of these two enzymes into cell walls(Fig. 4.4).

Abscissic Acid

Indirect effects

- Promotes ethylene production, hastens of senescence, and reduces auxin transport.

Direct effect

- Increases cellulase activity.

Summary of the Defoliation Process

1. Shift in the internal hormonal balance within the plant.
2. The abscission zone or abscission layer forms across the petiole near its junction with the stem.
3. Synthesis and secretion of the enzymes cellulases.

4. The enzymes digest the cell walls & middle lamella between certain cells of the abscission zone.
5. Cells on the stem side of the abscission layer expand, while they shrink on the leaf side. This causes the petiole to break off from the stem.

It is important to note that while the abscission zone consists of several layers of cells across the petiole, the vascular bundles remain intact, allowing water and nutrients to travel in and out of the leaf In practice, the water status of the plant is important. A small amount of water stress on the plant can reduce growth and facilitate defoliation to some extent. On the other hand, too much water stress on the plant can cause the leaf to desiccate, reducing the weight necessary to drop or break through the abscission layer. If defoliants are applied, they would have a difficult time penetrating the leaf due to a thicker cuticle and closed pores on the leaf surface due to water stress.

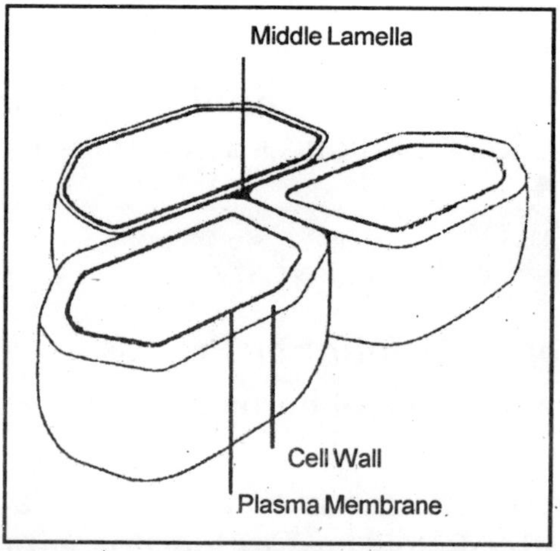

Figure 4.4. Enzymes digest the cell walls and the middle lamella of cells in the abscission zone, leading to leaf drop.

Environment
↓
Hormone Balance
↓
Pectinases and Cellulases in Abscission Zone
↓
Digestion of Cell Walls and Middle Lamella of Cells
↓
Unequal Growth, Pressure
↓
Leaf Drop

Figure 4.5. Summary of Defoliation Events

It is not essential that a cotton farmer develops a thorough understanding of the physiological or biochemical mechanisms associated with leaf abscission and plant senescence in order to accomplish a successful crop defoliation. However, as most farmers realize, the better someone understands how a piece of machinery works, the better chance they have in making it work correctly and to repair or correct any problems that develop. Similarly with crop management, the better we understand how a plant grows and functions, the better we can carry out operations such as cotton defoliation. As an act of controlled crop physiology, a successful defoliation takes place, in a balanced and directed process leading to leaf drop (Fig. 4.5). The combinations of crop management (plant-water relations, N fertility status, and boll load), environmental conditions, and the choice and timing of chemical defoliants serve to affect, and hopefully accelerate, the overall defoliation process.

5

Monitoring System for the Management of Cotton Growth and Fruiting

The life cycle of a cotton crop is the course of growth whereby plants develop from the seedling stage to the vegetative, reproductive, cutout, and finally, the maturation stage. Plant genetics and the environment interact to determine the time of initiation of each stage as well as its duration. Due to the indeterminate growth habit of cotton plants, the environment plays an important role in determining the time and duration of the various growth stages. For example, physiological and insect-induced abscission of reproductive structures reduces the competition for carbohydrates from reproductive organs, resulting in more vegetative growth & delayed cutout and maturity. Understanding the main physiological processes that occur during each growth stage is important in the efficient management of field crop production and yield. Surprisingly, the amount of time required by the cotton plant to pass through each stage of plant growth is quite consistent (Table 1). This is more or less fixed by the genetics of the plant and the environment.

Temperature is one easily measured environmental factor that influences plant growth. Plant physiologists have determined the amount of heat required for each stage of plant growth. The heat accumulated in a day is measured in Heat Units (HUs) or Day Degrees (DDs or DD60s). The ranges of accumulated HUs measured for different stages of cotton plant growth are presented in Table 1. Keeping track of accumulated HUs is a useful way to track the life cycle of a cotton crop to determine if it is proceeding normally.

Table 1: The average number of days and heat units (HUs) required by various growth stages in cotton (modified from Oosterhuis, 1992).

Growth stage	Number of days (range)	Heatunits (range)
Planting to seedling emergence	4-9	50-60
Emergence to first square	27-38	425-475
Square to white flower	20-25	300-350
Planting to first flower	60-70	775-850
White flower to open boll	45-66	850
Planting to cutout	80-100	1000-1600
Planting to harvest	130-170	2600
Between nodes :		
Up the mainstem	2-3	40-60
Out the branch	5-6	80-120

Efficient management of a cotton crop requires close monitoring of plant growth throughout the production season or life cycle of the plant. For the purposes of commercial crop management, the life cycle of a cotton crop can be divided into the following four phases: (1) Early Vegetative, plant emergence to appearance of the first square; (2) Juvenile, first square to first bloom; (3) Reproductive, first bloom to cutout; and (4) Maturation, cutout to maturation and harvest. The duration of the four life cycle phases in relation to the time course of plant height development and the time from emergence to first square, bloom and cutout. Each phase is characterized by a predominant physiological activity, and there are specific cultural practices that producers may use to optimize plant growth during the various phases of development. The objective of this management system is to achieve key developmental goals set for each phase of development. Management decisions are made by quantifying and evaluating plant growth, square and boll distribution and retention using plant maps. Plant mapping is a powerful technique used to assist cotton managers in tracking and quantifying, plant growth and yield potential throughout the various phases of crop development. These more exact techniques for monitoring and managing plant growth and development, can be a component of "precision farming" programs.

Several plant mapping techniques have been proposed for cotton (Bourland and Watson, 1990; Hake et al., 1991; Landivar et al., 1993; Zhang et al., 1994). Some of these programs use computers to summarize and display the status of plant growth and fruiting. The research group at the Texas A&M University Research and Extension Center, Corpus Christi, TX, has been involved in the development of a Plant Map Analysis Program for cotton, or PMAP (Landivar, 1993). PMAP is a computer program especially designed to facilitate plant map data entry and analysis. Data entry and editing are accomplished through the use of a spreadsheet template that resembles a cotton plant. The user can move to any fruiting position on the plant template displayed on the computer monitor screen by using a pointing device or the arrow keys on the keyboard. Once the data are entered into computer files, PMAP can display the data in text or graphic form. A unique characteristic of PMAP is its ability to compare plant maps collected from different experimental treatments or commercial fields. The comparisons are displayed graphically or in tabular form to facilitate data analysis and interpretation. While PMAP was developed for research purposes, the flexible approach used to enter and display data makes it suitable for use in the analysis of data collected for many other purposes, including the management of commercial cotton fields (Hopkins et al., 1995). Another plant monitoring program that is popular is COTMAN (Zhang et al., 1994). This program is oriented toward insect management as well as crop agronomic management. Based on results of plant monitoring COTMAN suggests management practices that may improve yield potential. For example, it provides a decision of when to terminate insecticide applications after cutout.

Phases of cotton Plant Development

Phase 1 - Early Vegetative Phase

It begins with seed germination and the unfolding of the cotyledons, continues through the formation of true leaves and ends with the appearance of the first square at position one on the first sympodial branch (i.e., fruiting branch). Depending on temperature, this phase may last between 27 and 38 days from the time the cotyledons are fully expanded until the appearance of the first square.

While above-ground plant growth is slow during this phase, plants are actively developing their root systems. A sluggish above-ground growth rate may be caused by low levels of solar radiation intercepted by the plant canopy and by cool air and soil temperatures that are normally encountered during early spring. McMichael (1990) presented data showing that the root-to-shoot ratio is approximately 0.35 at 12 days after planting and declines to 0.15 at 80 days after planting. He stated that in the develop-mental processes, both total shoot length and dry weight increase as the plant grows, and continue to increase until a maximum plant height is reached and the boll load begins to develop. Management objectives during this phase are to: maintain a uniform plant population, develop a vigorous root system, and produce healthy foliage and growing points.

The second goal to achieve during this phase of crop development is the production of a vigorous root system. McMichael (1990) recorded data showing an exponential root weight gain pattern from ten until forty days after planting. One of the most limiting factor to root growth during the early vegetative phase was shown to be low soil temperatures. Bland (1993) reported that cotton roots could grow downward at a rate of 0.7 inches per day in a soil profile uniformly warmed at planting to a temperature of 17.8°C and then progressively warmed to 30°C at 100 days after emergence. Treatments simulating a fast and slow warming of the soil profile decreased downward root growth to 0.6 and 0.3 inches per day, respectively. Under some cool adverse conditions, root growth may be further restrained by some herbicides incorporated in the top 3-6 cm of soil. Oosterhuis and Zhao (1994) recorded data from a greenhouse experiment showing that the application of the growth regulator PGR-IV significantly enhanced root and top growth. In a controlled environment study, the enhancement of root growth would presumably result in a subsequent increased top growth, by initiating a vigorous growth cycle of roots and shoots. Locke et al. (1994) presented data from a field experiment which has shown that the application of PGR-IV at planting and at the initiation of first squares resulted in increases in above ground plant dry weight at approximately 50 days after emergence. The application of PGR-IV and other plant hormone-based growth promoters may help to enhance early vegetative growth, particularly for early planted fields preceding favorable soil and air temperatures or in years with lower than normal temperatures during conventional planting dates.

The third goal to achieve during the first phase of crop development is the protection of the foliage and growing points against insects. Early insect infestations such as cotton aphids (*Aphis gossypii* Glover) and thrips (*Frankliniella* spp. and *Thrips* spp.) can cause a considerable reduction in leaf area, leading to reduced plant vigor.

In areas where aphids and thrips are chronic pests, the use of at-planting soil-applied, systemic insecticides can efficiently control infestations and prevent early damage. However, if aphids and thrips are rarely pests, then close monitoring for their presence and density and the use of foliar insecticides, if necessary, may be the most economical approach to control these pests and achieve the desired growth, plant structure, and square production. Cotton fleahopper and tarnished plant bug damage to the terminal growing point may also result in the outgrowth of several monopodial (vegetative) branches from the injured terminal. This unusual growth habit is commonly referred to as "crazy top." Close monitoring of cotton fleahopper and tarnished plant bug populations and use of insecticide sprays, if necessary, are essential toward the end of this phase.

Phase II - Juvenile Phase:

The juvenile phase of crop development begins with the appearance of the first visible square, continues throughout the initiation of sympodial branches and squares, and ends with the appearance of the first bloom on the plant. The duration of this phase is determined by temperature and usually lasts 25 to 35 days. The main objectives to accomplish during the juvenile phase of development are to: (1) obtain normal mainstem elongation and leaf area expansion rates, and (2) manage for good production and retention of flower buds.

Plants enter a linear phase of weight gain and mainstem elongation during this phase of growth. Good growing conditions during the early vegetative phase and, in particular, the juvenile phase seems to give cotton plants the potential to develop a full canopy surface area that is able to receive and use its maximum of incoming solar radiation before the initiation of the reproductive and boll-filling periods. Cotton plants begin producing the first fruiting branch (sympodial branch) at the fourth to sixth mainstem node above the cotyledonary nodes, depending on cultivar and temperature. The first four to six mainstem nodes are capable of producing only vegetative branches (monopodial branches). At temperatures of 22.2° to 25°C, plants begin to produce one new sympodial branch on the mainstem every 3 days. Temperatures below 22.2°C delay the time interval to more than 3 days between new sympodial branches. Conversely, temperatures above 25°C reduce the time interval to less than three days (Baker and Landivar, 1991; Hodges et al., 1993). By the appearance of the first bloom, plants have normally developed 14 to 16 mainstem nodes above the cotyledons. Fruiting sites along sympodial branches, that develop into squares, blooms, and bolls, are initiated every five to six days depending on temperature (Baker and Landivar, 1991; Hodges et al., 1993).

Cotton researchers agree that most square-shed prior to the development of the boll load and cutout may be insect related rather than physiologically related. Tarnished plant bugs and cotton fleahoppers can damage the terminal of reproductive branches, stopping the development of the branch. Over-wintered boll weevils (*Anthonomus grandis grandis* Boheman) can infest cotton fields and injure squares as they reach the matchhead size and larger. The use of insecticide applications at matchhead square are effective to control over-wintered boll weevil populations and moreover helps to suppress subsequent generations of boll weevils from injuring flower buds and bolls in phases three and four of crop production. Bollworms [*Helicoverpa zea* (Boddie)] and tobacco budworms [*Heliothis virescens* (F.)] can also injure squares and terminal growing points at this time. For the above reasons, the main management activities during the juvenile phase of crop development consist of monitoring and suppressing insect populations to prevent square and terminal damage.

Phase III - Reproductive Phase, Bloom to Cutout

This important phase of crop development begins at bloom, continues with the initial growth and maturing of the bolls, and ends at cutout with the fertilization of the last harvestable bolls. Phase three is between four and six weeks in length, depending on environmental conditions. The end of this phase coincides with the time in which the uppermost first position bloom is located five nodes below the terminal, and for practical purposes, the date of this event is defined as the date of cutout. Cutout is a key event or guidepost to use in management of cotton.

During this phase of development, plants continue to grow linearly in terms of mainstem elongation rate. Maximum plant height and light interception (canopy closure) are attained during this phase. The initiation of blooming, flower pollination (fertilization) and boll growth and maturing characterize this period in the production cycle. As the number of bolls on the plant increase, their potential capacity to utilize carbohydrates increases in direct proportion to their number. In contrast, the carbohydrate supply does not increase proportionally to the number of leaves, but approaches a ceiling i.e., asymptote set by the amount of light interception. As the plant continues to grow according to environmental conditions (temperature, light, water, nutrients), the capacity of the bolls to use carbohydrates will eventually exceed the supply, and the bolls will compete with each other and all other vegetative and reproductive tissues for carbohydrates. This event, where the bolls are using almost all available carbohydrates, has been recognized as the cause of cessation of growth and is commonly referred to as cutout. It normally occurs at 80 to 90 days after planting.

Cutout is affected by the growth habit of the particular cultivar, environmental limitations, and management practices. The fast fruiting habits of early maturing cultivars result in a high demand for carbohydrates by reproductive organs at a time when the canopy, root system and other vegetative organs are still growing and expanding. This often results in early cutout of a crop with incomplete canopy cover and reduced rates of photosynthesis per unit of land area. For these reasons, more determinate cultivars tend to cutout and mature earlier. Similarly, environmental constraints, such as drought, can reduce vegetative development (leaf area and root expansion) and photosynthesis, and induce premature cutout. Cutout can also be induced by management practices that increase fruit retention such as early insect control practices and the use of plant growth regulators.

Cutout dictates the end of the growing season that is, the end of production of new blooms and bolls that will contribute to harvest in areas where cotton is grown as an annual crop. For management purposes, the date of cutout is defined as the time at which the uppermost bloom is five nodes below the mainstem terminal. At this time, the last harvestable bolls are being pollinated. Bolls pollinated after cutout develop under intense competition with older bolls for carbohydrates and nutrients, as well as under increasing insect pressure. For these reasons, late set (late pollinated) bolls have reduced chances of reaching maturity, and a large proportion will be immature, insect-injured, or have fallen off the plant by harvest. The crop has reached its potential yield at the end of this phase, and from this point on, the best the farm manager can usually do is to maintain the existing yield potential. Management objectives during this period are to: (1) obtain optimum final plant height, (2) minimize square and boll abscission and promote boll maturity, and (3) induce timely cutout.

The mainstem elongation rate proceeds linearly at early bloom, and therefore, height control continues to be an important objective to accomplish. Final plant height should be approximately 10 to 20% greater than the row spacing. High retention of first and second position squares and bolls often creates sufficient competition for carbohydrates to induce delays in plant development. These delays can stop the initiation and elongation of mainstem nodes and internodes and increase abscission rates of squares and bolls.

As bolls develop, the need for nutrients and moisture increases. Providing adequate nutrition to the developing bolls is often confounded by reduced root activity due to the competition for carbohydrates from squares and bolls. If the root system is not capable of providing the necessary nutrients, foliar fertilization should be considered. Application could prevent or reduce square and boll abscission and insure optimum boll growth. The nitrogen requirements of a highly productive cotton crop may be as high as 1.4-1.8 kg per acre per day; therefore, frequent applications of supplemental amounts of nitrogen (N)may be needed in order to obtain a yield response to foliar applications. A single foliar application of 2.27 kg or less of N per acre during this period of intense boll development is relatively ineffective.

Phase IV - Maturation Phase

The final phase of the crop production cycle begins at cutout and ends with the application of defoliants and boll openers. This phase may last four to six weeks, depending on fruit load, water supply, temperature, and soil nutrient availability. At the initiation of this phase, boll growth continues at a high rate and a large number of harvestable green bolls less than 20 days old will continue to be susceptible to insect damage. Vegetative growth during this phase has completely ceased due to carbohydrate demand by the bolls and decrease in photosynthetic ability of the canopy as the leaves age. Management objectives during this phase of crop growth are to: (1) protect the last harvestable bolls from insect damage, (2) determine when to terminate insecticide applications used to protect the bolls, and (3) determine the proper timing for the application of defoliants and boll openers. Determining the date of cutout is essential in estimating the time for the termination of insecticide applications (Bernhardt et al., 1986). Studies have shown that bolls reaching 14 to 20 days of age (i.e., 350 to 450 HUs) are far less susceptible to injury from boll weevils, and bollworms and tobacco budworms less than 6 days old, than younger bolls (Bagwell, 1994). Therefore, some researchers have recommended termination of insecticide applications for these insects at about 20 days after cutout (i.e., cutout plus 350 to 450 HUs) (O'Leary et al. 1996; Cochran et al., 1996). The basis for this thinking is that at 20 days after cutout all harvestable bolls will be 20 days old or older and relatively safe from boll weevil, bollworm, and tobacco budworm injury.

Plant Mapping to Evaluate Crop Performance

The plant mapping analysis consists of sampling fields at key stages of development. Several programs are available, such as PMAP (available from Juan Landivar) and COTMAN (available from Cotton Incorporated). Use of PMAP will be discussed first.

The recommended sampling method for PMAP is to divide a field into management units according to soil type, planting date, cultivar, etc. Limit the size of a management unit to approximately 100 acres; or it may be larger than 100 acres if a field is uniform. A minimum of 32 plants should be mapped from 6 or 8 randomly selected sites. Care should be taken to select sites with a plant population equal to the average plant population of the field being evaluated. The plant mapping procedure suggested here, using PMAP, requires recording the type of reproductive structure present (i.e., square, boll or fruit missing) at the two first fruiting positions on each reproductive branch as well as measurements of plant height and number of vegetative nodes. At bloom, it takes approximately one hour to map 32 plants and 15 minutes to enter the data into computer files. Plant mapping should be performed at the following times during crop production: (1) early juvenile phase, (2) early reproductive phase, (3) late reproductive phase, (4) maturation phase, and (5) an optional sampling prior to final harvest.

First Mapping at Early Juvenile Phase

The first plant mapping is recommended at approximately 10 to 15 days after the initiation of the juvenile phase (first square). At this time, plants may have initiated 10 to 12 mainstem nodes above the cotyledons. Objectives of this sampling are to evaluate the percentage of square retention and plant vigor. Seventy-five to eighty-five percent retention of the first and second position squares have been subjectively rated as normal. Levels above or below those percentages are rated as high or low square retention, respectively. Low levels of square retention at this time may be a warning for the need to scrutinize early-season insect management in order to avoid further losses of squares. It may also be the first indication of a potential for rank growth, and subsequent plant map data collected during the early reproductive phase should confirm this potential for rank growth. The first fruiting branch should be at four to six nodes above the cotyledon. Cotton crops that initiate the first fruiting branch at or above node six on the mainstem have a tendency toward developing a large vegetative mass and delayed maturity. PMAP generates a frequency distribution of the number of squares and mainstem nodes for the plants sampled. This frequency distribution is useful in determining the uniformity of the field and in quantifying the fraction of the plants that are significantly behind or ahead of the average development for the crop.

Second Mapping at Early Reproductive Phase

The second plant mapping is done at 7 to 14 days after the appearance of the first bloom. At this time, each plant should have 2 to 5 green bolls and have initiated approximately 12 to 15 reproductive nodes on the mainstem. In addition, plants have also initiated most of the reproductive structures that will contribute to final lint yield. The objective of this mapping sample is to verify the efficacy of early insect control and to estimate fruit retention and potential lint yield. Crops with 60 to 70% retention of first and second position structures (squares, blooms, and bolls) are rated as having an average yield potential. Crops with retention levels below or above this level are rated as having a yield potential below or above the historical average yield, respectively. Although final lint yield is still heavily dependent on subsequent weather conditions, the retention level of reproductive structures at this phase of development is helpful data in estimating the possible pay-off of management alternatives. Fruit retention levels obtained during this phase of development should confirm the potential for rank growth and delayed maturity of the crop. Table 2 provides management guidelines based on the level of fruit retention measured at the first two fruiting positions on each sympodial branch (i.e., positions one and two).

Table 2: Interpretation of levels of fruit retention obtained during the early reproductive phase of development (second plant mapping sampling). Fruit retention of 60 to 70% are considered average

	Fruit retention at first and second position	
Factors Effected	Below 60%	Above 70%
Yield potential	Below average	Above average
Potential for rank growth	Higher	Lower
Need for Pix	Higher	Lower
Need for nutrients	Lower	Higher

Third Mapping at Late Reproductive Phase

The third plant mapping should be collected during the third or fourth week of bloom (20 to 25 days after first bloom). This mapping may provide the most important information to crop managers. The objective here is to determine when the fruiting position with the last harvestable bolls were pollinated (estimated cutout date) and to obtain an update on final yield potential. The date when the last harvestable fruit were pollinated is important, because it allows the estimation of the number of insecticide applications that would be required to protect young bolls from insects such as boll weevils. This information is generated by analyzing the distribution of the number of nodes above the uppermost bloom or green boll for each plant sampled.

The number and distribution of green bolls provide important information on the potential yield of the crop. PMAP estimates yield potential using a predetermined potential yield contribution of each fruiting site on the plant. The unit for this measurement is the number of bolls at a given position needed to produce a gram of lint. The potential lint yield per unit of area is obtained by multiplying the number of bolls at a given position needed to produce a gram of lint by the fraction present at each position. The sum of these products is the potential yield for the crop. This information along with the data on the number of mainstem nodes above the highest bloom can be used to determine the cost/benefit of subsequent insecticide treatments. The data provided from this mapping also provides information to estimate crop maturity data for defoliation. This estimation is made by adding 30 to 40 days to the date when the uppermost bloom is located five nodes below the terminal. The fourth plant mapping sample will confirm the timing for the termination of the crop.

Fourth Mapping at Early Boll Opening or Maturation Phase

The fourth plant mapping should be performed approximately one week after the appearance of the first open boll. The objective of this sampling is to accurately estimate the time to apply chemical defoliators and to evaluate the potential consequences of untimely applications of defoliants. Defoliation timing can be determined by analyzing the distribution of green and open bolls by reproductive branch.

Fifth Plant Mapping - Harvest (Optional)

The PMAP program estimates the values of each open boll and produces a plot displaying the economic value of each reproductive branch.

Monitoring using COTMAN

COTMAN is a cotton crop information system that records changes in the fruiting dynamics of the cotton plant as well as in plant growth parameters that are useful as a prompter of timely management decisions. The utility of COTMAN in describing the progress of a cotton crop will not be fully realized until such information improves decision making. We hypthesize that such decisions often will require consideration of the balance between square abscission (reproductive) and plant main-stem nodal growth (vegetative), but questions remain about our ability to measure this balance and to effectively use the information. This system can be used to evaluate methods: (1) for detecting shifts in the balance between square abscission and plant growth, and (2) to determine if detection of the shifts could be used to improve management decisions.

In attempts to vary the balance between square abscission and plant growth management practices used were control levels of square protection, use of plant growth regulators (PGRs) or by varying plant stand density. Irrigation also is an important factor involved in varying the balance between fruit numbers and size, but irrigation has to be held constant in this experiment by furrow irrigation all treatments.

Treatments consisted of simulated insect damage, three plant densities, and two PGRs. The split-split plot experiment are designed to provide varying levels and types of plant stress. With four replications, main plot treatments were 2 levels of square abscission: protected as needed with insecticides vs damaged squares and sub-plot treatments. Three levels of PGRs applied, no regulators vs mesiquat include 3 levels of plant stand density.

The Retention/Growth Balance (RGB) equation is : RGB – $(X_2\text{-}Y_2 - X_1\text{-}Y_1)/(X_2\text{-}X_1)$, where X_1 and X_2 are the number of squaring nodes at two consecutive sampling dates, and Y_1 and Y_2 are the sqaare shed rate at two consecutive sampling dates.

The patterns of each growth curve compared to the target development curve show clear early evidence that we can detect stress due to plant density. The research can also confirm that the cotton crop can tolerate a high rate of square shed without undue yield loss. COTMAN indices reflect shifts in plant stress and strain indications. The aggregate change in retention-growth balance is a sensitive indicator of stress, and can be exploited in timely management decisions. Evaluation of management practices to avoid cotton plant stress seem possible with these standardized indices.

Mid-Season Plant Mapping

Cotton plant mapping after first bloom is a tool that can be used in making management decisions. Two new concepts have been introduced for this stage of plant mapping : Nodes Above White Bloom (NAWB) and Nodes Below White Bloom (NBWB). These two measurements are used to summarize growth (NAWB) and boll status (NBWB).

Sampling for mid-season plant mapping is similar to that for early season plant mapping. Randomly select at least 20 plants from throughout the field. Avoid plants adjacent to large skips and plants with damaged terminals. In addition, sample only plants with a first position bloom. Plants can be mapped in the field or after removal by cutting below the first fruiting branch. Because only nodes and first position aborts are recorded, the time for mid-season mapping is less than for early season mapping.

Nodes above white bloom (NAWB) indicates energy a plant has above that required to fill bolls, reflecting excess energy that produces additional terminal growth. Nodes above white bloom actually measures terminal growth between the appearance of a pinhead square and bloom, If the boll load uses most of the nutrients provided by roots and leaves, very little is left over for continued terminal growth, indicated by lower NAWB. When the boll load is low enough that excess nutrients are available, NAWB will remain high or possibly increase.

When mapping NAWB, begin counting at the first node above the first position white bloom and continue to the terminal node with a main stem leaf at least 1 inch in diameter. Count and record all the first position aborts (damaged or missing squares) above the white bloom.

Normally fields will have from 9 to 11 nodes above the white bloom at the early bloom stage. At mid-bloom this number generally drops to 8 or 9. If the number of nodes remains 9-11 at mid-bloom, then boll size or fruit retention is inadequate, and an application of the plant growth regulator PIX should be considered. If NAWB nears 5 at mid-bloom, there is a danger of early cutout and the crop was not pushed adequately with nitrogen and water.

Nodes Below White Bloom (NBWB) measures the fruiting progression. During the bloom period, heat units are generally more stable and, therefore, not as limiting to growth as in early or late-season growth periods; consequently, development from square to bloom occurs at a fairly consistent rate. An additional NBWB should accumulate every 3 days, allowing NBWB to follow crop progress through bloom. Medium to short growing seasons have a limited duration to effectively set bolls. Longer seasons may have the crop set on upto 15 nodes.

Using the same plants used for determining NAWB, count the number of nodes below the white bloom down to the first fruiting branch. Count and record all the first position aborts below the white bloom. This measurement divided by the NBWB will provide percent boll retention.

Boll retention is generally high at early bloom, but drops off toward cutout. Assuming a normal cotton crop is set on 10 nodes or less. NBWB provides an estimate of how much more season is necessary to set the crop. Cotton experiences its greatest shed during the young boll stage, as young bolls are particularly sensitive to physiological causes of shed such as water stress, high temperature, leaf damage, and excessive shading. When boll retention is low following high pre-bloom square set, it may need prompt attention to remedy the effects of physiological cause of shed.

In season plant mapping is a powerful cotton management tool. Observing the growth and development of the plant and its fruiting pattern closely, along with understanding what is normal, provides the cotton producer with appropriate options. Many times a poor situation can be substantially improved as a result of plant mapping. Consider the data obtained by mapping at harvest time. The first field had relatively low boll retention (less than 50%) on all fruiting branches. The second had 50% or better retention on nodes 9 through 14; the third, good boll retention on nodes 6 through 17. As would be expected, the yield of field first was relative low, field second was good, and field third was excellent.

Early season and mid-season plant mapping would have indicated problems with field one that could have been improved with the proper management practices. The cause for poor fruit set and retention could have been determined and appropriate measures to encourage better retention applied. The second field set and retained a good early crop. If the season was favourable, yield could have been improved by adding water and nitrogen to encourage additional fruit retention in the upper part of the plant. It is evident that conditions were favourable in the third field by the good boll retention throughout the plant. In season plant mapping showed what was happening in the field and allowed the producer to manage the crop, taking advantage of its yield potential.

□□□

6

Water Relations in Cotton

Cotton plants must maintain a balance between the water supply stored in the soil system and the atmospheric evaporative demand in order to carry out growth and development processes necessary for productivity and economic yield. In order for a plant to grow, it must carry on photosynthesis, which means it must absorb radiant energy and atmospheric CO_2. In doing so, energy is available to evaporate liquid water from the cell wall to water vapour in the intercellular spaces, which can easily diffuse out of the leaf into the bulk air when the stomata are open to allow CO_2 in for photosynthesis. As water vapour escapes from the leaf, liquid water is evaporated and the tissue water potential declines creating a gradient for water to flow from the soil to the leaf. The rate of vapour transport is called transpiration, and it has a direct effect on the resultant tissue water status, which can affect metabolic activity and growth processes. The soil represents the reservoir from which the plant roots extract water. Both texture and depth affect the water holding capacity of the soil system. When the soil water supply is not adequate to meet the transpirational demands, then plant water deficit stress occurs and growth and productivity are affected negatively. Maintaining the soil water supply at non-limiting levels is dependent upon very timely rains or the use of supplemental irrigation.

Potential Evaporation

Weather parameters of incident radiation, air temperature, relative humidity, and wind speed can be measured hourly or daily and used to calculate the potential evaporation rate of the atmosphere. The modified-Penman equation is currently the most widely used and accurate predictive model for estimating potential evaporation from cotton crop.

$$ET_o = c[W - R_n + (I - w) - f(u) - (e_a - e_d)]$$

where ET_o = potential crop evapotranspiration (mm/day), W = temperature-related weighting factor, R_n = net radiation in equivalent evaporation (mm/day), f (u) = wind related function, $(e_a - e_d)$ = vapour pressure deficit of the air at mean air temperature, and c = adjustment factor to compensate for the effect of day and night weather conditions.

The major source of energy required to convert liquid water to vapour is solar energy, the same energy that is required to produce chemical energy for the reduction of CO_2 to organic compounds in the photosynthetic process. The heat of vapourization of water (2.43 MJ kg^{-1} or 540 cal g^{-1}) is one of the highest for liquids. During the cotton growing season, total incident radiation levels in excess of 25 MJ day^{-1} are common. If all this energy were absorbed and converted to heat and used to evaporate water, the evaporation potential would be in excess of 10 mm day^{-1} or 0.4

inches. However, some of the incident radiation is reflected back into the atmosphere (albedo~25%) and some is re-radiated as long wave thermal energy (proportional to K^4), reducing the amount of energy available for evaporation.

The rate at which water vapour diffuses out of the leaf (transpiration rate) is a function of the vapour pressure deficit between the leaf mesophyll air space and the bulk air surrounding the leaf and the extent of the leaf surface conductance, which is dependent on stomatal pore size.

Transpiration Rate (mol H_2O $m^{-2}s^{-1}$*)* = $(C_{wv}leaf - C_{wv}air)$ *stomatal conductance:* where C_{wv} is the water vapour content. The vapour pressure of the air surrounding the leaf varies with changing temperature during the course of the day. The relative humidity changes as air temperature changes, declining as temperature increases. The dewpoint, which is a measure of the absolute water vapour content of the air, remains fairly constant during the course of a day. The air inside the leaf is very close to saturation (i.e., relative humidity is greater than 99%). The vapour pressure gradient between the leaf and the bulk air surrounding it steepens as the air temperature increases. Therefore, the actual rate of water loss from the leaf is dependent upon the vapour pressure deficit between the air inside the leaf and the air surrounding the leaf.

Crop Water Use

Crop water use rate (daily water loss) is a function of the daily potential evaporation rate times a crop coefficient (K_c), which is defined as the effective leaf area index of the crop. The ratio of actual daily crop water use to potential evaporation increases linearly with increasing leaf area index (LAI) reaching a maximum (1.0) at leaf area indices around 3. The slope of the relationship between LAI and K_c is affected by canopy structure and plant spacing.

The leaf surface is covered with a cuticle of water impervious substances such as waxes and oils. The stomata offer the only path for water vapour to escape the leaf interior. Stomata exist on both the upper and lower leaf surfaces of cotton. The average stomatal density is approximately 10,000 cm^{-2}. Stomatal conductance is highly variable and responds to radiation level, CO_2 concentration, atmospheric humidity, and leaf water status. The controls of stomatal activity consist of both physical and biochemical components. When the soil water supply is non-limiting, stomatal conductance is at its maximum dependent upon the physical conditions. When soil water supply becomes limiting, stomatal conductance declines to restrict water loss.

As liquid water is converted to water vapour inside the leaf and it moves out of the leaf, the liquid water concentration declines in the cellular tissue, establishing a water potential gradient within the entire plant-to-soil liquid water continuum. Water potential ($Ø_w$) is a measure of the free energy status of water in solution and is the driving force for water movement from a region of high free energy ($Ø_w$ close to 0) to a region of low free energy ($Ø_w$ = negative). The water potential of the leaf tissue is dynamic and reflects the changes in evaporative demand and soil water supply on a daily basis. When soil water potential is high (close to field capacity), the leaf water potential reaches a minimum value dependent upon the cellular osmotic potential ($Ø_n$). In most cotton plants this minimum $Ø_w$ is around - 1.3 Mpa. When the soil water potential ($Ø_s$) becomes limiting (approximately -0.2 Mpa) then as evaporative demand increases, either the leaf water potential must continue to decline ($Ø_w$ declines) or stomatal conductance must decrease to reduce water loss to maintain a favourable tissue water status.

Soil Water Supply and Water Deficits

Soil water supply is dependent on soil texture and depth. The texture determines the water holding capacity per unit depth. As clay content increases, the water holding capacity increases due to increased surface area per unit volume of soil and due to the surface electrical charge associated with clay particles. Cardinal water contents have been established for each soil texture and consist of field capacity and permanent wilting point values. The extractable soil water between field capacity and permanent wilting point is defined as plant available water.

Cotton plants begin to suffer tissue water stress long before the soil water content reaches permanent wilting point. The quantity of water a plant can extract from a given soil before suffering stress depends on root length density and evaporative demand conditions. Plants with high root length densities (cm roots cm^{-3} soil) or plants experiencing relatively low evaporative demand conditions can extract a greater amount of the plant available water before suffering stress than plants with lower root length density or that exist in higher evaporative demand conditions. Cotton has a relatively low root length density when compared with other crop species. Therefore, cotton plants begin to suffer water stress at considerably higher soil water potentials than do other crop species. Under very high evaporative demand conditions, cotton plants can suffer short-term water stresses on a daily basis even though the soil water supply is close to field capacity. When tissue water stress is experienced, either stomatal closure must occur to restrict water loss or the tissue must adjust either cell size or cellular osmotic potential to allow the cell water potential to decline to maintain the flow of liquid water. Stomatal conductance reductions to limit water loss also can represent limitations to CO_2 diffiusion into the leaf to support photosynthetic activity. Stomatal conductance and CO_2 assimilation rate change in parallel with decreasing leaf water potential resulting in reduced growth rates.

The period from square initiation to first flower represents the most critical developmental period in terms of water supply affecting yield components. Both boll number (per plant) and boll size have fairly strong positive correlations with water supply during this period of development. Water supply from first flower to peak bloom also affects boll number. However, it appears that the production of fruiting sites rather than retention of young fruit is more critical to yield and responsive to water supply under short growing season environment. Water stress prior to flowering reduces fruiting site number due to inhibition of site initiation rather than due to square shed. Stress after flowering affects young fruit abortion. Both the supply of reduced carbon and nitrogen as well as a hormonal imbalance in the young fruit are responsible for fruit abortion. It appears that young fruit less than 10 days of age are most vulnerable to abortion.

When rainfall is not adequate to maintain the soil water supply above the level where plant stress begins to occur, then irrigation can be successfully used to supplement the rain. Two considerations must exist for successful irrigation application to occur. The rate of crop water use must be known and the replacement capacity of the irrigation system must be known. The rate of crop water use has been discussed above and consists of knowing the potential evaporation rate of the atmosphere and multiplying that by the crop coefficient at that time. The best measure of irrigation capacity is gallons per minute per acre (GPMA). Each GPMA is equivalent to 0.052 acre in. per day of replacement capacity. If the cotton crop is using 0.25 in. per day (PET x K_c) then an irrigation capacity of 5 GPMA is required to provide the daily water use.

The total water supply to the crop (stored soil water + rain + irrigation) must be maintained at adequate levels to reduce the risk of plant water stress occurring from the onset of squaring until the last harvestable fruit is 10-12 days old. Then water supply must be managed to allow plant stress to develop to reduce the retention of fruit that have little chance of being harvested and to allow maturation of retained fruit.

How irrigation water is applied has a major influence on both the application efficiency (per cent of delivered water that actually can be used by the crop) and the distribution efficiency (uniformity of application across the area). Flood irrigation has the poorest distribution efficiency. Sprinkler irrigation has better distribution efficiency than flood methods, but has low application efficiency due to wetting the entire soil surface allowing for a relatively high percentage of the applied water to be lost to evaporation from the bare soil surface. LEPA (Low Energy Precision Application) is a system that uses a center pivot adapted to drop the liquid water on the soil surface (eliminating evaporation from the nozzle to the ground). Circular rows are required for maximum efficiency. Water is delivered to alternate rows, wetting less than 50% of the soil surface and increasing the per cent applied water stored for crop usage to over 90%. This application system is most effective when the water supply is not adequate to meet the evaporative demand of the crop. Not only is the delivery system important, but also how frequently can water be applied. Using the LEPA system, a 3-5 day frequency produces the highest yields compared to longer irrigation intervals. Caution must be used at very low application volumes to reduce the evaporation percentage. It is recommended that no less than 0.5 in. be applied at a time. The disadvantage of the LEPA system is the very high application rates per unit of land area. The outside irrigation spans are applying water at rates that greatly exceed the infiltration rate of even the sandiest soils. In order to reduce runoff, furrow dikes are strongly encouraged.

□□□

7

Environmental Stress and Cotton Growth

Abiotic factors are the most important as they affect crop growth and yield. Crop production is directly influenced by temperature, photoperiod, total radiation, and precipitation. Temperature determines the duration of a crop's growing season. Some species are sensitive to photoperiod and will not flower in extremely long days or flower prematurely when exposed to an inappropriate photoperiod relative to its stage of growth. However, modern cotton cultivars are not photoperiod sensitive even though their ancestors were. Temperature controls cotton development and indirectly its water requirement. Total radiation is an important environmental factor, as it directly drives crop photosynthesis, growth, and indirectly controls all plant processes.

Long-term averages of weather parameters are defined as the climatology of those factors for a region. Although each crop is controlled by the physical environment to which it is exposed, one can learn whether a particular crop is suitable for a given location or region by studying the long-term weather records.

Water is critically important, but good quantitative knowledge of the relationships among water supply, other weather factors, and plant water status is not available. Therefore, cotton plants growing in the natural environment are often prevented from expressing their full genetic potential for yields. Attempts are made to optimize moisture conditions by controlling amounts and timing of water application. High yields are the result of growing crops in a high radiation environment and correcting the most limiting environmental constraint, water deficit.

TEMPERATURE

Optimum temperature for cotton production is 27.7°C. However, long-term average daily mean temperatures can exceed 30°C for 88 days per year. Although such high temperatures would be expected to prevent commercial production of cotton, evaporation from the leaves of well-watered plants has an important cooling effect on the crop canopy. This is particularly true in a desert environment where the humidity is low and evaporation is high. There is a linear relationship between leaf water potential and leaf temperature minus air temperature. The relationship between leaf water potential and the difference between leaf and air temperature shows that as water deficit stress becomes more severe (-2.0 to -3.0 MPa), the leaf temperature becomes warmer. This indicates that plants close their stomata in stress conditions. Such a process conserves the moisture within the plant when the soil water supply is limited, but also results in warmer leaves causing modification and, in extreme cases, heat damage to various growth processes.

Though average daily temperatures in midsummer exceed the optimum temperature for cotton growth, producers can manage to successfully grow cotton by planting early. By planting early, the crop escapes the midsummer temperature extremes.

Growing degree days (GDD's) are sometimes used to predict crop growth and developmental events, because cotton growth increases linearly as temperature increases between 15.5°C and 31.1°C. Since cotton does not grow at 15°C or cooler, growing degree days 60 (GDD_{60}) are calculated from the positive difference between the average daily temperature and the base temperature (15.5° C).

$$GDD = \frac{Tmax, ^{\circ}C + Tmin, ^{\circ}C}{2} - 15.5\ ^{\circ}C$$

where GDD60 is growing degree day accumulation for a day with maximum (Tmax), minimum (Tmin), and the base temperatures. The GDD60 can be added to and accumulated as the season progresses from planting or day of emergence. Growing degree days are a useful way of summarizing temperature effects on crop growth and development. Temperature is the most limiting variable for the crop much of the season, and the crop developmental as well as growth responses are quite linear. If conditions are dry or nutrients are limiting, then crops do not respond directly to temperature in a predictable manner like they do in nonlimiting conditions, and GDD's accumulate without the usual advance in growth and development. Therefore, GDD's are not useful for predicting crop development in stress environments. This becomes particularly important in midsummer when other environmental stresses frequently limit crop growth. More information is needed to understand crop development in stress conditions.

Short-term Temperature Variability

Daily and yearly temperature variability is another environmental factor that impacts crop growth and development. Long-term average daily temperature provides some understanding of the average conditions, but individual years obviously vary. Planting is usually determined by temperature and rain. Seedling establishment and the prevalence of early-season diseases and insects are strongly influenced by temperature. Below-optimum temperatures cause slow cotton growth and favor pest development. After flowering begins, high temperature may cause flower retention problems. Therefore, growers must manage their crops each year and deal with the unique combination of conditions provided by nature.

Diurnal Patterns in Temperatures

The uniqueness of individual days is relative to optimum or minimum temperatures at which cotton grows. It should be remembered that environmental conditions determine growth and survival of organs on individual days, and ultimately determine yield. Extreme conditions for a few hours will cause vulnerable tissues or organs to die and subsequently abscise from the plant. In addition, at low but non-limiting cool temperatures below 15.5°C, cotton metabolism is so slow that essentially no growth or development occurs and result in abscission of many young bolls. Such abscissions reduce yield, delay maturity, and favor rank growth.

Temperature and Photosynthesis

The photosynthetic response of plants to increased temperature and atmospheric CO_2 is quite similar to vegetative-growth responses. The response of photosynthesis to increasing CO_2 was smaller at the lowest (17.2°C) and at the highest temperatures (36°C) compared to the responses at temperatures more favorable for cotton growth. The optimum temperature for vegetative structures, particularly leaf growth, in both pima and upland cotton is slightly higher than the optimum temperature for producing fruiting structures. The higher photosynthetic rates at both ambient CO_2 and twice ambient CO_2 are likely because of the continued addition of new leaves at those high temperatures. Previous work also found that temperatures above optimum for fruiting result in fruit abscission and, therefore, less intra-plant competition for carbohydrates. Normally cotton, being an indeterminate-flowering type plant, slows the addition of leaves and fruiting sites as the boll load increases, resulting in what is commonly called "cutout." As bolls mature and no longer require nutrition, photosynthetic products accumulate and encourage vegetative growth commonly known as "re-growth." If the bolls prematurely abscise, the intra-plant competition for nutrients does not occur, and more new leaves continue to be produced throughout the fruiting period. Bolls may abscise prematurely because of injury by insects, nutritional deficits, or injury by environmental factors.

Young cotton leaves reach their maximum photosynthetic rates about 14 to 16 days after unfolding, then the photosynthetic rates of individual leaves gradually decline until leaf senescence. Therefore, the relatively higher rate of net photosynthesis that occurred at 26.7°C or higher compared to that of plants grown at 17.2 °C was probably because of the younger population of leaves in the canopy of the high-temperature grown plants. More leaves were initiated on both the mainstem and branches at high temperatures than at low temperatures. Leaves also expanded more rapidly at 30°C than at either 20 to 36°C, resulting in a younger leaf population in a canopy. One would expect therefore, that in a warmer, high CO_2 climate, photosynthetic rates of plants will be greater because of both a steeper CO_2 gradient from the atmosphere to the CO_2 fixation sites inside the leaves and the younger, more efficient leaf population on canopies produced in that environment.

Temperature and respiration

Cotton plants can store only very small quantities of sugars. These are therefore quickly converted to starch or translocated to growing organs anywhere in the plant. Each growing organ may use that sugar to support growth or some "work" activity. Approximately 50% of the sugar each growing organ receives is used as building blocks to form structural material, i.e., that portion of the sugar supply becomes incorporated into the cell walls as cellulose or other structural compounds. The remaining 50% of the sugar that arrives in a growing organ is used as an energy source for making the chemical conversions essential to create proteins, lipids, or other structures used to build cells; or it is used to do the other necessary work. Examples of the other work include providing the energy for mineral uptake by the roots, or transporting sugars and other nutrients throughout the plant. Energy for these activities, building new molecules and structures and providing the biochemical energy for other work, is obtained from a process called respiration.

Respiration is sensitive to temperature. As temperature increases so does respiration. Therefore, plants growing in a high-temperature environment require more sugars for respiration than similar plants in lower temperatures. Plants that are stressed because of inadequate water may also become warmer as their stomata are closed due to water loss. An indirect effect of the water deficit then causes a greater need for sugars because of the higher temperatures; a lower sugar supply also results because closed stomates reduce photosynthesis. A deficiency of leaf nitrogen also quickly results in a shortage of sugars required for respiratory-supported growth and other work. Plants stressed because of water or nutrient deficits, or plants simply stressed because of high temperatures, may have insufficient sugars to support all their structures. This may result in the death of tissues that were previously actively growing.

Temperature and Early-season Vegetative Growth

The temperature control can be set to maintain several constant variants from ambient conditions. This includes plants grown at all growth stages at temperatures that were real relative to natural conditions, and ± variants from that. As temperature increased, the time required to produce squares decreased in a linear manner, as did flowering or open bolls. The rates of the processes (time to first square, first flower or first mature boll-time when lint can be seen through the cracked sutures) were different, but the responses to temperature were linear.

Temperatures during seedling establishment played an important role in cotton growth. The plant height of cotton plants growing at 21.1°C was about 1 inch at 20 days after emergence, while plants growing at 27.8°C averaged 7.6 inches tall and those growing at 30°C were over 9 inches in height. There were small but not significant differences in the heights of plants in different CO_2 Concentrations at the warm temperatures, but essentially no height differences in plants were limited by low temperature. Mainstem node formation in response to temperature was essentially the same as plant height. Dry matter (above ground) accumulation and leaf area at 20 days after emergence also increased similarly to plant height as temperature increased. There was a greater response of dry matter accumulation to elevated atmospheric CO2 and the dry matter accumulation response was greater in plants growing at near optimum temperature.

Temperature and Balance between Vegetative and Reproductive Growth

Large numbers of squares were produced at optimum temperature. More were produced at the high temperatures than at the lower temperatures. Since the bolls abscised from plants growing at those high temperatures, and continued to produce new nodes and fruiting sites. The number of bolls retained was very strongly controlled by temperature regimes in which the plants were grown. There was a distinct advantage to growing plants at the higher temperatures early in the season. At 20 days, the high-temperature-grown plants were 4 to 7 times larger than plants grown in optimal temperatures. However, that advantage was lost because of failure of the high-temperature-grown plants to retain fruit.

Seedling responses to warmer temperatures typically occur during seedling establishment period. The high-temperature-grown plants produced large numbers of squares, and those squares matured to produce flowers. The squares did not abscise or abscised in only very small numbers. The very young bolls (3 to 5 days after anthesis) abscised in the higher temperature regimes. Irrigation during flowering can lower canopy temperatures and reduce high-temperature injury.

The total vegetative weight (leaves, stems, and roots) produced by plants grown at the various temperature regimes do not differ. Plants grown in high CO2 environments produced about 40% more vegetative mass than plants grown in ambient CO2 environments. However, the plants grown in the high temperatures matured earlier and were harvested earlier than those grown in the cooler environments. Therefore, the rate (mass per day) of vegetative dry matter production was higher in the high temperatures.

At the lower temperatures, doubling the atmospheric CO2 consistently increased boll mass in small amount, but in the two higher temperature regimes, the boll mass was insignificant because of the abscission of young bolls. The small mass values are due to green bolls that were initiated late in the season when the temperatures declined sufficiently.

Temperature and Boll Retention

The number of bolls produced was not influenced much by different temperature regimes. However, the number retained was strongly controlled by temperature. When plants were grown at 29.8°C, all the flowers set for the first 5 days were retained. After that, the percentage retained gradually declined as the boll load on the plants increased. Bolls produced and retained are also presented as a function of days after temperature treatment It is clear that there is a very low probability of cotton production at such high temperatures in a humid environment where evaporation from leaves is low.

Fruit production efficiency, defined as dry weight of fruit per dry weight total mass produced, increased as average daily temperatures increased to 29.4°C, then it declined rapidly as temperature increased above 29.4°C. Since the rate of fruit retention dropped so dramatically at temperatures above 29.4°C. The calculated fruit-production efficiency also dropped at those high temperatures. The survival and growth of bolls are not equally sensitive to high temperature throughout their development. Bolls usually abscised within 2 to 4 days after flowering when exposed to high temperatures. It has been observed that when the temperature was not damaging for a few days, the flowers produced during that period survived and bolls grew to maturity even though temperatures were above 32.2 °C for several days during the fruit-growth period.

Apparently, there is a short period associated with flowering when the reproductive process is most vulnerable to average daily temperatures (above 32.8 to 34.4°C). If the fruit escapes high temperature during that time, it can survive unfavorable, high-temperature conditions during the rest of its growth period. However, an exact definition of the vulnerable period is lacking. Such high temperatures cause rapid maturation of the bolls so that less time is available for seed and lint growth. Thus, even if the bolls escape average daily temperatures above 29.4 °C during flowering, which causes boll abscission, exposure to such high temperatures later in the boll growth period will cause smaller bolls. When reciprocal crosses were made between plants growing at different temperatures, i.e., in some cases, pollen from high-temperature-grown plants was used to pollinate flowers produced at optimum temperatures, and, in other cases, pollen from optimum-temperature-grown plants was used to pollinate flowers produced at high temperatures. The high temperatures appeared to injure both pollen and the female flower parts, and the injury appeared to be progressive with time of exposure. It appears that fertilization and seed set were damaged by high temperature 10 to 12 days prior to flowering. Increasing CO_2 did not ameliorate the heat-sensitive condition in cotton.

Breeding both high- and low-temperature-tolerant cultivars will be beneficial in a future warmer world. Low-temperature tolerance in cotton would allow producers to plant the crop earlier and permit a crop canopy to develop earlier in the growing season. This would allow flowering to occur before the midsummer high temperatures that limit fruit set. The damaging effect of high temperature on cotton production has been recognized. Cotton cultivars tolerant to high temperature transpired more water than high-temperature-sensitive cultivars. The high-temperature-tolerant cultivars had cooler canopies because they transpired more; thus they essentially escaped the damaging high temperatures.

Temperature and Boll Development

Developmental events occur much more rapidly as temperature increases. Number of days to the appearance of first square, first flower, and mature open boll decreased as the average temperature to the respective events increased. The numbers of days required to produce squares, flowers, and mature fruit were reduced by 0.9, 1.7, and 3.8 days, respectively, per degree increase in temperature. Thus, assuming the temperature increase will be equally distributed throughout the growing season, a 12.8°C increase in average global temperature will speed development from emergence to maturity by 24 days. Unfortunately, most of the shortening of developmental time occurs during the boll growth period and results in smaller bolls. Bolls that mature rapidly because of high temperature have less time available to grow and reach their genetic potential size. Therefore, in years that are extremely hot during the flowering and boll-filling period, the crop matures earlier and has smaller bolls than in years in which the temperature during that period is cooler.

An examination of the GDD60 results illustrates the effects of seasonal changes in temperatures on the catch-up of crop developmental events when planted at different times. There was a difference of 42 days in emergence dates of these crops, but if the average GDD60 occurred the dates of squaring were separated by only 15 days. We often see such a "catching up" of crops planted when the temperature is more favorable for growth and development. Many flowers were produced on plants growing at plus 12.8 or 10.6°C, but the fruits abscised soon after flowering. There was no loss of squares due to high temperature in any of the environments. More fruiting sites, and thus more squares and flowers, were produced in the high- CO_2 and high-temperature-grown plants. Doubling the atmospheric CO_2 did not affect the developmental rates or protect the plants from injury to the fruit-set process at high temperatures. Thus, cotton plants developed fruiting sites more rapidly in temperatures above 30°C, but the young fruits abscised because of such high temperatures. As a result, well-fertilized and well-watered crops continued producing additional flowers and leaves and maintained a canopy capable of rapid photosynthesis while in hot temperatures (average daily above 30°C). Variation in temperature in the natural environment allows some flowers to escape exposure to damaging temperatures so that some bolls are eventually produced. Unfortunately, long growing seasons resulting from such a scenario are expensive to sustain because of the long period the crop which must be provided with nutrients, water, and protection from pests.

Temperature and yield variability

The stagnating and fluctuating yields relate to extremes in the weather conditions, period when boll development peaks. The problem relates to genotypic changes over the last many decades. Traditionally, farmers think of yield in terms of number and weight of bolls per acre. More realistically, the two fundamental components of yield are the number of seeds produced per acre and the weight of fiber on those seeds. While management and environmental factors influence the former, the latter is determined primarily by genetic considerations, and secondarily by environment. Since cotton production's ultimate objective is fiber, interest must focus on fiber weight per seed and number of seeds per acre.

Although cotton originated in hot climates, it does not necessarily yield best at excessively high temperatures. Growth chamber studies indicate 20 to 30 degrees Celsius (C) is cotton's optimum temperature. From a physiological perspective, its best temperature range for optimal metabolical activity is 23.3 to 32.2°C, with 27.8 being the very best temperature for photosynthesis. When temperatures exceed 33 degrees, growth rate declines. Ironically, average daily maximum temperatures in July and August peak boll development months almost always exceed 32.2°C, well above the photosynthetic optimum. A positive relationship between yield and temperature during boll development, illustrated that high temperatures correlate with low yields, and cooler temperatures correlate with high yields. These correlations were strongest in August, but also applied to a lesser extent in July. Above-average day time temperatures reduce photosynthesis and carbohydrate production. There is no sharp threshold, but rather gradual decline in net photosynthesis, with a greater than 50 per cent decrease at about 33°C.

Night/Day Degrees Matter

Under normal conditions, a single cottonseed produces about 12,000 to 18,000 fibers. High temperatures and the resulting decreased carbohydrate availability can reduce boll size by decreasing both number of seeds per boll and number of fibers per seed. High temperatures can affect pollen formation and subsequent fertilization, resulting in fewer seeds per boll. Apparently, cotton has an upper temperature limit, above which the yield development rate begins to decrease, although this is not well defined. Sustained, excessively warm nights above approximately 23.3°C significantly increased respiration, causing additional carbohydrate loss. The overall result of high temperature is that available carbohydrates are reduced and may not be sufficient to satisfy all plant needs, which reflected in increased boll shedding, malformed bolls (e.g. parrot beak), smaller boll size and lower yields. High temperature is a particularly relevant factor when boll loading is its maximum developmental phase.

Cotton fiber consists mainly of carbohydrate. Thus, decreased carbohydrate availability is manifested in less fiber per seed and lower gin turnout. Comparing the two production systems reveals that the main reason for the large yield difference is that during boll development, when nighttime temperatures average about 13.3°C cooler than optimal night temperature. Therefore, cotton does not experience the increased respiratory losses that have been shown to be detrimental to optimum boll growth.

Moisture and High Temperature

Water deficit worsens high temperature's detrimental effect on growth, and, unfortunately, low rainfall periods often coincide with hot weather spells. Normally, cotton tries to regulate plant tissue temperatures by the cooling process of water evaporation through numerous stomates, which are also important for permitting carbon dioxide (CO_2) to enter for photosynthesis. When water is available, the evaporative process (approximately 540 calories per cubic centimeter of water evaporated) can keep leaves 15.6 to 14.4°C below air temperature, causing the leaves to "feel" cool when touched. However, during a drought, stomates close, the evaporative cooling process stops and leaf gets heated up above the optimum temperature range for photosynthesis and carbohydrate production. Closed stomates also limit the CO_2 supply. So, when rainfall is in short supply, dryland cotton is more sensitive to high temperatures than irrigated cotton.

Alleviating Heat Stress

Generally, variable and low yields during hot years cannot be blamed on poor management. There are, however, some obvious strategies to adopt. Genetic improvement is being pursued including selection in breeding lines for temperature tolerance, and also screening new and yield type germplasm for tolerance. Management options should focus on producing an early crop, not delaying flowering and early boll development into the hottest, driest period. Using crop monitoring techniques (e.g. COTMAN) to prevent stress and square and boll dynamics should also help to ensure timely management inputs and an early crop. Plant growth regulators should be used to enhance early fruit set and early maturity. Recent evidence indicates certain plant growth regulators might help during mild stress by improving carbohydrate translocation to developing bolls. One innovative idea being investigated is the application of an osmolyte spray such as glycine betaine or proline to improve heat and/or drought tolerance.

Despite best management efforts, untimely, severe weather can still adversely affect growth yield. Current research seeks to understand what happens during boll-filling and devise methods to alleviate problems that disrupt this crucial process. This includes breeding for temperature tolerance, using crop monitoring, using earlier-maturing varieties to escape the hottest, driest part of the season, irrigating at appropriate times and so forth.

Water-deficit stress

Water deficit worsens high temperature's detrimental effect on growth, and, unfortunately, low rainfall periods often coincide with hot weather spells. Normally, cotton tries to regulate plant tissue temperatures by the cooling process of water evaporation through numerous small leaf pores known as stomates, which are also important for permitting carbon dioxide (CO_2) to enter for photosynthesis. When water is available, the evaporative process (approximately 540 calories per cubic centimeter of water evaporated) can keep leaves 4ºF to 6°F below air temperature. However, during drought, stomates close, the evaporative cooling process stops and leaves heat above the optimum temperature range for photosynthesis and carbohydrate production. Closed stomates also limit the CO_2 supply. So, when rainfall is in short supply, dryland cotton is more sensitive to high temperatures than irrigated cotton. Generally, variable and low yields during hot years cannot be blamed on poor management. There are, however, some obvious strategies to adopt. Genetic improvement is being pursued including selection in breeding lines for temperature tolerance, and also screening germplasm for tolerance.

Timely irrigation supplements to achieve optimum moisture conditions can be regulated by three evaluations. First, examine soil moisture and plant stress indices to determine need. Second, take evapotranspiration and soil moisture measurements. Third, consult infrared stress graphics to determine the need for supplemental watering. By placing irrigation scheduling on an established seven-week schedule with watering occurring each week, maximum lint yield can be expected. Increasing moisture each week by about 0.5 inch (4 inches in seven weeks) could increase yields. Providing moisture evenly across fields and minimizing moisture loss from evaporation heightens irrigation efficiency.

In cotton, moisture makes or breaks fiber quality. Water stress below -22 or -23 bars (a bar is a unit of measuring dryness based on the pressure equal to one million dynes per square centimeter), particularly in late-bloom stages, will reduce late-developing bolls and fiber strength in mid-canopy bolls and increase micronaire of existing bolls. Cotton fiber length and strength and even seed weight and oil percentage are governed by moisture. Fiber length is most affected at 16 to 20 days after flowering. Fiber strength is most affected 25 to 30 days into boll development through three to four days prior to boll opening. During extreme drought, sudden irrigation right at or just after the peak in flowering will reduce lint percentage. Sudden good moisture during boll development will encourage seed weight rather than lint weight, although fiber length and micronaire value may be improved.

Cultivar choice also is important. Full-season varieties with an indeterminate growth type allow plants to recover from drought stints. Pima and Acala types may demand more moisture than Upland varieties. Full-season varieties often are the best-yielding varieties in stressful situations, as the longer growth period allows more time to adjust, shift and regain yield. Adapted, full-season cotton varieties can be a good choice on too well-drained soils if planted in a timely manner. Drought-tolerant genetics or even slight, timely preconditioning on crop plants also can aid in providing sustainable yields even during drought periods. Unadapted, short-season cultivars may cut out or have reduced recovery ability due to genetics and limited-season growth. Only under severe drought conditions with complete crop loss will little difference be seen across crop maturities. Occasionally, in extreme drought, a short-season variety can develop and yield a small amount as compared to complete losses with later maturing varieties. Generally, however, the full-season varieties adapt better to seasonal differences.

Water stress and nutrients

Sufficient nutrient levels help cotton progress through water shortages. However, some nutrients are more limited in uptake during drought. Nitrogen and potassium uptake is limited during drought stress. Supplying extra nitrogen can compound problems late in the season, when rain or irrigation between cut out and harvest cause cotton regrowth or defoliation problems. Indeed, if drought is forecast, it may be wise to use an average, not excessive, fertilizer allotment for an average yield. If moisture conditions improve by midseason, petiole testing and foliar nutrient applications can boost nutrients and, perhaps, yield.

Water stress and Pests

Pest management to reduce drought damage is very important. Early seedling diseases in cotton not only weaken plants but also limit moisture extraction from soils. Drought actually can reduce budworm, bollworm, aphid and borer pressures in cotton fields but armyworms, spider mites and grasshoppers can reach alarming numbers. Scouting is even more essential during droughty cropping seasons.

Water status by Remote-Sensing

It is possible to monitor subtle changes in crop water status by measuring changes in crop reflectance with a ground radiometer. Most importantly for irrigation scheduling purposes, these changes were apparent before the irrigation was due. During the five day period prior to the irrigation date or refill point, reflectance decreased in the near infrared and green bands. Red reflectance increased only after the soil moisture status fell below the refill point. The greatest change in reflectance was in the near infrared region, which was attributed to a decrease in GC% caused by canopy architectural changes including leaf wilting. The experiments showed that the crop reflectance was more strongly influenced by ground cover and plant height than the soil moisture status.

Potential crop yield, conventionally assessed as fruits counts, were not related to the reflectance data. The risk of using remote sensing technologies to predict yield is that the sensors respond to crop parameters such as leaf area, and not actual fruit numbers. Although advanced canopy development is indicative of high potential yield, poor insect control or unchecked proliferation of foliage negate this relationship. The effects of leaf senescence and appearance of white cotton bolls on crop reflectance indicate these remote sensing tools could be effective in monitoring crop defoliation.

Crop temperatures rose with soil moisture depletion and were influenced by soil moisture availability and soil compaction. Thermal infrared crop temperatures offer the most potential for irrigation scheduling, although the data is strongly influenced by soil background, for the first half of the cotton season. Finally, to monitor crop moisture status the thermal infrared temperatures offers the most potential, whilst to monitor crop canopy growth (ground cover, LAI, height) the near infrared wavelength would be most suitable.

Water stress and cut-out

Droughts may make cotton cut out early or it may extend early growth cycles in plants, depending on the stress and timing of the dry periods. Ultimately, drought usually causes cotton to open early. If cotton fruiting is compressed, maturity and defoliation are affected. Early crops may need to be defoliated sooner. Begin examining and opening a few bolls to determine defoliation timing once 40 per cent of the bolls are open in drought years. If boll development is close to the target defoliation time, schedule and apply defoliation materials to ensure a quick and efficient harvest. In cases where residual nitrogen may promote late-season regrowth, desiccant use also may be needed.

Water stress and Carbohydrate production

Production of carbohydrates through the process of photosynthesis and their storage are the primary functions of leaves. Leaf age is an important plant factor affecting daily photosynthesis. In non-stressed plants, peak carbohydrate production from an individual leaf occurs when the leaf is around 20 days old. Peak plant carbohydrate production will occur when the combination of photosynthesis per unit leaf area and leaf area is maximized. In non-stressed plants this usually occurs some 60 to 70 days from the unfolding of the first true leaf (75 to 85 days after planting). Decline in daily carbohydrate production after this date results from increasing canopy leaf age and increased self shading and the increase in boll demand for carbohydrates, which restricts any new leaf development. Water stress has been shown to reduce whole plant leaf area largely through reductions in total leaf numbers. The rate of leaf expansion is also reduced, which in turn reduces the size of individual leaves.

Reduction in leaf area will obviously affect the level of total canopy photosynthesis. Photosynthesis is maintained in priority over leaf expansion and development. This allows the plant to maintain current photosynthetic capacity but limits future capacity. It can be seen that leaf expansion has been greatly reduced by the time photosynthesis has started to decline. In terms of plant growth, the maintenance of photosynthesis will enable boll and root growth to continue longer during periods of water stress than vegetative growth. This drought adaptation will also allow the plant to recover quickly from small periods of mild stress, particularly during early, pre-flower growth stages.

Carbohydrate redistribution

The export or redistribution of carbohydrate from an individual leaf is initially small, to allow effective leaf growth. Once leaf growth has stopped at around

20 days of age, however, carbohydrate in excess to leaf requirements is produced and this can either be stored as starch for later use or be directly exported to actively growing plant parts.

The rate of export of carbohydrate from the leaves is determined by the demand imposed by other plant parts. Actively growing organs such as roots, bolls and growing terminals act as 'sinks' which actively compete for the available carbohydrate. The pattern of distribution will depend on the leaves, capacity to satisfy the requirements of individual sinks. Since, it is particularly important for the plant to have the capacity to utilize excess carbohydrate during periods of stress. Plant adaptation allows the processes involved in transferring carbohydrates away from the leaves to continue at higher water stress levels than those that reduce photosynthesis. Therefore, water stress not only affects production of carbohydrate but also alters its distribution.

Root development

At the time of flowering, around 80% of the plant's root system may be developed and thus, the root system imposes the greatest demand for excess carbohydrates during early plant growth. After flowering, boll development begins to compete with the root system for carbohydrates and the rate of root expansion declines. Under water stress, however, the plant is adapted to place priority on root growth. As a result, root expansion occurs at the expense of vegetative and boll growth.

Leaf development

Leaf expansion is sensitive to water stress. Drought inhibits cotton canopy development, it decreases the number of leaves on the sympodial branches. Leaf area of glasshouse-grown cotton was inhibited when the percentage of soil-available water was less e.g. leaf water potential below -.5 bars decreased leaf elongation rate. After two days of withholding water, the leaf expansion of 55-days old plants slowed showing that leaf growth was more sensitive than root elongation to drought. The terminal drought decreased shoot : root ratio.

Leaf abscission

Abscission of leaves from cotton plants is induced by an unfavourable plant water balance. The amount of abscission is directly related to the magnitude of internal water deficit. An association between ethylene production rate of intact cotton petioles and plant water deficit has been presented. Ethylene participates in water-stress induced abscission. Changes are initiated at plant water potentials near -10 bar which pre-dispose cotyledonary leaves to the abscission incuding action of ethylene. Atleast two types of changes account for this response, the ethylene production rates of the tissue may increase as a function of the internal plant water deficit, or the tissue sensitivity to ethylene is increased during the development of internal water deficits.

Ultrastructural changes in leaves

Water-stress induced marked alteration of ultrastructural organization of both chloroplasts and mitochondria. Ultrastructural organization of chloroplasts was sometimes almost completely destroyed; chloroplasts and ribosomes disappeared but peroxisomes seemed not to be affected. Accompanying water stress was a sharp increase in activity of acid phosphatase, acid and alkaline lipase (glycerol ester hydrolase) within chloroplasts. Only acid lipase activity was detected inside mitochondria. These alterations in cell organization account for reported effects of water stress on CO_2 compensation point, photochemical reactions and photorespiration.

The effects of water stress on photosystem-II (PSII) fluorescence and photosynthetic electron transport activities were analysed in cotton. The collection of photons, and their distribution between two photosystems, and PS II photochemistry are unaffected by the stress. The results have shown drought slows down the rate of plastiquinone reoxidation and PS I mediated electron transport is inhibited.

Boll development

Squares exhibit little carbohydrate demand on the plant during early growth, with bracts supplying the majority of their requirements. A rapid increase in demand for carbohydrates occurs after flowering .This is the reason that the majority of fruit is shed as flowers or two or three day old bolls. Shedding of bolls can occur upto an age of 10 to 14 days, after which cell wall thickening between the boll and stem prevents an abscission layer from forming. In the case of a rapid onset of water stress, young bolls in which growth has stopped may be retained by the plant and appear as 'mummified' dry bolls. The boll development is affected by total carbohydrate supply and not by the rate of distribution from adjacent leaves. This is consistent with the fact that redistribution of carbohydrate can occur at stress levels beyond those that affect production.

As boll demand exceeds supply from the adjacent leaf, inter-boll competition for further carbohydrate occurs. Older bolls compete more effectively than younger bolls and this results in the movement of carbohydrates away from the extremities of the main stem and individual fruiting branches. Those bolls unable to compete effectively are either shed by the plant or are reduced in size, hence the occurrence of smaller bolls towards the top of the plant. It is for this reason that the majority of fruit, particularly secondary position bolls, is retained by lower fruiting branches.

In non-stressed irrigated crops, increased early vegetative growth results in shading of lower leaves and this causes reduced retention and boll size on the first two or three fruiting branches. The final results of this combination of inter-boll competition and leaf shading in fully irrigated crops is the common bell-shaped distribution of bolls throughout the plant. In the case of crops under water stress, the same inter-boll competition occurs, but there is generally less total carbohydrate to be distributed amongst bolls. Reduced vegetative growth also minimizes shading of lower leaves, resulting in the higher boll retention and boll size occurring amongst the first fruiting branches.

Crop water use and plant growth

A cotton crop's requirement for water changes throughout the growing season, following the pattern of evapotranspiration. The rate of evapotranspiration is determined primarily by meteorological factors and the availability of soil water. Total crop evapotranspiration will vary with canopy size, or leaf area. Crop leaf area peaks 3 to 5 weeks after the start of flowering and this results in a peak in daily water use between 8 and 10 mm.Maximum demand for water also coincides with the growth period between peak flowering and early boll development. Exposing the plant to water stress at this stage of growth can result in significant yield reductions. The impact of water stress at different crop growth stages on final yield is directly related to the water demands expressed by the crop. Stress during periods of high water demand can produce large reductions in yield. Stress during peak flowering can double yield losses compared with early or late seasonal stress. The impact of any one stress period is increased if followed by further stress.

Water stress and crop yields

The degree of plant response to stress will vary depending on the level of stress which occurs and the timing at which the stress is imposed, relative to crop growth. summaries the plant's responses to differing degrees of water stress.

Fibre development

Under water stress younger bolls are shed to enable the development of older bolls. The plant has increased its adaptation for survival during drought by placing priority on seed and fiber development over total fruit retention. This is demonstrated by the fact that young boll shedding can occur at lower moisture stress levels, while fiber development is not affected until higher stress levels are reached. The increase in micronaire generally associated with cotton suffering from water stress at the end of flowering is a good example of this plant adaptation. Increase in micronaire occurs because younger bolls are shed, and more carbohydrates becomes available to lower bolls. With fibre development continuing under higher stress levels, any extra carbohydrate available is allocated to increases in fibre cell wall thickening, leading to increases in micronaire. Moisture stress during peak flowering will tend to affect fibre length rather than fibre maturity, while stress later in the season will primarily affect fibre maturity.

Even though cotton is a drought-tolerant crop, it responds well to sufficient water by producing lint proportional to amounts of rainfall or irrigation supplied. Timely rainfall early in the season will establish stands and in the summer months will sustain squaring, flowering and bolls even beyond that normal 40 to 50 per cent of fruiting structures. Fertility, cultivation, cultivar choice, cultural practice, pest management, defoliation treatment, timely harvest, residue management and timely and adequate supplemental irrigations all work to optimize cotton production during drought periods. The three key periods of cotton growth that should be supplemented with moisture occur at stand establishment, prebloom and shortly after boll set. Ample moisture at seeding will establish healthy, uniform stands. Ample moisture before the first bloom will provide water for pollination processes. Ample moisture between initial boll opening and 50 per cent boll opening can provide key reserves for fiber strength and length development.

Severe drought on cotton plants will slow plant development and cause small bolls and squares to shed. Establishment and prebloom irrigations affect total yield, but water deprivation following bloom and into boll development also affects lint quality.

If managed properly, cotton can withstand drought on infrequently irrigated, coarse-textured, sandy soil with hot, dry conditions. Even during peak bloom, cotton will use only about 0.3 to 0.4 inch of water per day. Irrigated cotton fields, of course, respond well to additional moisture with yields escalating as moisture needs are met. Research shows that timely, supplemental irrigation can bring in more lint per acre in cotton.

Water stress can restrict both vegetative and boll growth. It has been shown that no matter what degree of water stress is imposed on a crop, the proportionality between vegetative growth and boll development remains relatively constant. Similar results have been achieved with crops receiving different amounts of nitrogen. This implies that, independent of water or nutrient supply, the plant will always attempt to form a balance between vegetative growth and boll development.

Table: Summary of responses to water stress

Degree of water stress	Possible causes	Physiological plant responses	Yield effects on maturity and WUE
Minimal stress	Reducing irrigation deficit	Excessive vegetative growth	Reduced yield
	Excessive rainfall	Increase in leaf area	Reduced boll size
	Cloudy weather	Extended flowering cycle	Delayed maturity
	Excessive early insect damage	Reduced carbohydrate surplus for bolls	Normal fibre length but low micronaire
	High plant stands	Reduced root development	Poor WUE
		High boll capacity but poor boll retention	
Mild stress	Optimum irrigation deficit	Optimum vegetative growth rate	Maximum yield
	Average temperatures (not excessively hot)	Leaf expansion restricted	High quality cotton
		Photosynthesis remains unaffected	No delay in maturity
		Maximum carbohydrate surplus	Maximum WUE
		Maximum boll development	
		Good fibre development	
Moderate stress	Increased irrigation deficit	Reduced vegetative growth and leaf expansion	Reduced yield
	Extremely hot temperatures with low humidity, windy conditions	Reduced photosynthesis	Early maturity
	Little cloud cover	Reduced surplus carbohydrates	Increased short fibre micronaire
		Reduced boll carrying capacity	Slight decrease in WUE
		Increased fibre development	Slight decrease in WUE

Contd..

Table: Contd..

Severe stress	Less than 3 irrigations	Vegetative growth greatly reduced-stops after flowering	Low yields
	Dryland crops	Greatly reduced carrying capacity	Short fibre
		Little surplus carbohydrates	High or low micronaire depending on stress pattern
		Low boll retention	WUE depends on rainfall

Radiation stress

Radiation is an important factor affecting crop photosynthesis, development, and yield. Low photosynthetic photon flux density (PPFD) in cotton canopies is usually associated with adverse weather and excessive vegetative growth. Shade during flowering and fruiting significantly decrease the photosynthetic rate of cotton leaves and increased fruit abscission, resulting in low lint yield and poor fiber quality. Cotton plants under low PPFD during reproductive growth, fruited more slowly and had less fruit retention than plants grown under normal PPFD.

Cotton fruit abscission is closely associated with the supply of carbohydrates, mineral nutrient status and plant hormone balance. The carbon metabolism of plants is extremely sensitive to change in environmental conditions, especially light. Some studies have shown that low PPFD considerably decreased non structural carbohydrate concentration in cotton plants. However, boll shedding was not commonly related to the carbohydrate or N levels in cotton plants since no differences in the carbohydrate and N levels existed between plants with variable rates of shedding. The retention of bolls at the first fruiting position was greater than the retention of bolls at the second position, although the concentrations of soluble sugar in their excised receptacles and ovaries were similar, and therefore, differences in assimilate levels did not explain the variation in boll retention among cotton genotypes.

The shade during the reproductive growth period significantly decreased leaf net photosynthetic rate of field-grown cotton, resulting in considerable reduction in non structural carbohydrate (especially starch and sucrose) concentrations of the leaves and bracts. Decreased leaf photosynthetic rate of shaded cotton was primarily attributed to factors other than stomatal conductance. The total non-structure carbohydrates (TNC) concentrations of floral buds were not affected by shade during squaring and early flowering, but were significantly decreased by shade at the boll development (BD) stage because of increased boll load. The negative effect of shade on non structural carbohydrate concentration of floral buds increased with plant aging and distal fruiting position. Thus under low PPFD insufficient photo-assimilate supply during fruiting is a major factor causing increased fruit abscission, and decreased lint yield and fiber quality of field-grown cotton.

□□□

8

Nutrition

Cotton takes up nutrients as cations and anions from the water held between soil particles - the soil solution. The plant must expend a considerable amount of energy to take up nutrients. Nutrients are normally in much greater concentration in the plant tissue than in the soil solution. Only a small fraction of the total nutrient content of the soil is found in the soil solution. As nutrients are removed from the soil solution, they are replaced from labile forms held within the soil, which include organic matter ,low solubility minerals,cation and anion exchange sites on clay particles and organic matter

Processes that release nutrients into the soil solution are usually reversible. Where high concentrations of a nutrient exist in the soil solution (especially after fertilizer application) that nutrient may be precipitated as an insoluble mineral, making it unavailable to the crop until the soil solution becomes depleted.

The rhizosphere

The rhizosphere is the zone of soil surrounding the root where soil microorganisms flourish in great abundance, relative to the rest of the soil. Microorganisms proliferate here due to exudation of nutrients, sugars and other materials from the root. The rhizosphere is a zone of intense biological activity and cycling of nutrients.

Long fallow disorder of cotton is associated with poor mycorrhizal colonisation, since long periods of bare, weed-free fallows or growth of non-mycorrhizal crops reduce the amount of vesicular arbuscular mycorrhizae (VAM) in the soil. The decline in VAM fungi has serious implications for rain grown cotton production, where it is important to fallow fields to store moisture. Inoculation with VAM is not successful. The best means of keeping VAM active in the soil is to keep crops growing in a rotation system with short fallows.

Absorption of Nutrients

For nutrients to be absorbed by plant roots from the soil solution, they must come in contact with the plant root surface either by:

Root interception: As roots grow through the soil, they intercept and absorb the nutrient ions they encounter.

Mass Flow: Water moves through the soil to the plant roots as the plant transpires. The nutrients dissolved in the soil water are carried to the root surface as a result of the mass flow of water to the plant root. Nutrients that are abundant in the soil solution (eg Ca^{2+} and Mg^{2+}) may be carried to the root by mass flow in sufficient quantities to satisfy the crop requirement. In contrast, the P concentration in the soil solution is very low, so mass flow contributes a minimal amount of the P required by the crop.

Nutrient Uptake

Nutrient uptake may be restricted where a deficiency of one nutrient will reduce the uptake of other nutrients by limiting crop growth. The nutrient uptake slows as the crop matures and are cycled from vegetative to reproductive organs within the plant. Oxygen is required to maintain metabolic processes including nutrient uptake, hence waterlogging will slow nutrient uptake.

The poor soil physical structure i.e. compaction or chemical toxicities in the soil eg salinity or sodicity may limit root and shoot growth and nutrient uptake, even where sufficient nutrients are available. In cotton farming systems, deficiencies of N, P, K, S, Mg, B, Zn, Fe and Cu have been identified.

Nutrient mobility with in plant

The cotton plant can take up nutrients quickly, as demand requires. The highest demand period for most nutrients occurs from flowering to boll fill (ie during the period of fastest growth). Nutrients are stored in leaf tissue and other organs until required by the developing bolls. Storage is important to provide nutrients to the crop in periods when crop uptake is reduced (eg cloudy weather, waterlogging episodes). This is especially significant for N, P and K where the supply of these nutrients may not be able to meet the demand eg potassium at boll fill and the premature senescence syndrome. Nutrients vary in their mobility within a plant. Some nutrients have very low mobility and deficiency is seen in new growth.

The largest pool of most nutrients is the soil organic matter. Biological activity results in the release of nutrients (mineralisation), which are available for uptake by a crop. The crop and soil microorganisms themselves require nutrients, and convert some mineral nutrients back into organic forms (immobilisation). The soil organic matter comprises a variety of living organisms (eg plant roots, animals, bacteria, fungi etc) and dead or decaying material (eg crop stubble, humus) from which mineral nutrients are released. Humus is composed of organic material derived from plant and animal cells. It is the end point of microbial activity and is stable and relatively resistant to further biological decomposition. Because the mineralisation process is biological, it is affected by temperature and water conditions e.g.low temperatures and dry soil slow the mineralisation and immobilisation processes. These processes operate concurrently, cycling nutrients between the organic and mineral pools.

Organic matter chelates nutrients in the soil

Chelates are complexes of organic matter and metal ions such as zinc, iron, manganese and copper. Chelates can help maintain the supply of these nutrients in a form available to plants. If unchelated, these ions would quickly precipitate out of the soil solution to form insoluble minerals that are not available for crop uptake.

Clay minerals and organic matter have both positive and negative charges on their surfaces, enabling them to attract and retain both positive ions (cations or metallic ions) and negative ions (anions or non-metallic ions). Humus retains mineral nutrient ions to its charged surfaces in similar fashion to clay particles. It contributes to the cation exchange capacity (CEC) of a soil, although clay will contribute substantially more to the total CEC of the soil. Soils normally contain more negative than positive charges. Anions and cations with more than one charge (+ or-) are retained more strongly by the soil.

Cation exchange

Cation exchange capacity (CEC) is the ability of soil constituents to attract and retain cations H^+, Al^{3+}, K^+, Na^+, Ca^{2+} and Mg^{2+}. Soils having high clay and organic matter contents will possess higher CEC than sandy soils of low organic matter content. Anion exchange also takes place on the clay minerals and organic matter (as with CEC) but anions are attracted and retained to the positively charged sites. Anions include nitrate (NO_3^-), chloride (Cl^-), phosphate and sulfate. Anions with single charge (eg nitrate, sulfate and chloride) are more prone to leaching down the soil profile, whereas anions with multiple charges (eg phosphate PO_4^{3-}) strongly resist leaching. Base saturation is defined as the total CEC occupied by the cations K^+, Na^+, Ca^{2+} and Mg^{2+}. The base saturation reflects the extent of weathering and leaching that has occurred in the soil. Base saturation is related to soil pH and is an indicator of soil fertility. The availability of the nutrient cations K^+, Na^+, Ca^{2+}, Mg^{2+} to the crop increased with higher base saturation. Soil pH governs the equilibrium of chemical reactions in the soil. The availability of nutrients to the plant is also affected by soil pH. The application of fertilizers affect the soil pH in the fertilizer zone, and can temporarily alter micronutrient availability. As ammonia is nitrified to nitrate, the pH in the band may fall to neutral (pH 7). This change in soil pH following the application of nitrogenous fertilizer may improve the availability of some micronutrients.

Nutrient requirement depends on variety and location under both irrigated and rain fed conditions. *G arboreum* requires less nutrients than *G.hirustum* and hybrids require more than single varieties. The short compact variety types are more efficient users of nutrients than tall late varieties. Due to extensive conditions under which cotton is produced (climate, varieties Soil), it is very difficult to generalize recommendations for cotton fertility management.

N increases height of plants, their boll-bearing capacity ,seed weight, and improves seed cotton yield. Excess N - e.g. if cotton follows a good crop of soybeans - may result in rank growth, poor fruiting, delayed maturity, difficult defoliation and boll rot.The requirement of N is less for Asiatic cotton types than for American types, but it is greater for hybrids. The response to N is better in irrigated than in rainfed cotton. For better N response - as well as for other nutrients - proper weed control, insect and disease control is essential. To obtain the highest efficiency from N, one-half or one-third should be given at the time of sowing, and the remainder split between square formation and peak flowering and boll development. Split applications (side-dressing) will better meet the crop's requirements and reduce nitrate loss.N given before or at planting can be broadcast together with P and K and incorporated into the soil; but placement at a depth of

5 cm below and 5 cm away from the plant has given a better yield than broadcasting or simple drilling. It is also common to apply part of the N as a foliar spray (1.5 - 2.0 % urea solution) in combination with insecticides at the time of peak flowering and the boll development stage.

Foliar application has been found most beneficial on rainfed cotton.

P improves root development, water use efficiency, the energy balance and the weight, oil and protein contents of the seed as well as fibre quality. The response to P (as well as to K)is normally better if it is given in form of NPK fertilizer (e.g. with N:P2O5:K2O ratio 2:1:1 or 3:1:1), before or at planting, incorporated into the soil or placed 7.5 - 10 cm below the moist soil surface. When not applied pre-planting, P should be given at the latest at thinning (in the bottom of the furrow before watering).

K improves fibre fineness and strength, and results in early maturity of the crop. The response to K will always be better if it is applied with N and P at preplanting and incorporated into the soil. Only on sandy, permeable soils it is appropriate to give a split application of K, e.g. with a second dressing at the time of thinning.

Mg: If soil status is low apply 45 - 50 kg/ha MgO; where liming is recommended, dolomitic limestone may be used.

S: If the applied NPK fertilizer does not contain S, at least 12 - 15 kg/ha S should be given in addition (e.g. in ammonium sulphate).

Mn: On soils with pH above 5.6, application of 2.8 kg/ha elemental Mn is recommended.

Zn: Soils are deficient when Zn drops below 0.55 ppm; in this case 4.5 kg/ha of 36 % zinc sulphate should be given possibly as foliar spray with insecticides.

B: where needed, apply 0.6 kg/ha B if possible in two foliar applications at 0.3 kg/ha each, tankmixed with insecticide spray

Liming the crop is very sensitive to soil acidity. It has repeatedly shown a marked response to lime when applied to an acid soil, particularly if pH has dropped below 5.5 approximately.

NITROGEN

The plant use and efficiency of fertilizer N is a function of time of application, rate of the N applied, and precipitation and climate-related variables. The maximum fertilizer N efficiency can be related to the latest application being compatible with the stage of crop development associated with maximum uptake. Therefore, information pertaining to crop N requirements e.g. amount of N needed to produce a given unit of yield and the uptake and utilization patterns for the crop in question are considered as fundamental to developing N management strategies that optimize uptake and efficiency. With respect to cotton fertilization a nutrient balance approach to N management provided the best results in terms of fertilizer uptake and recovery in both irrigated and dryland conditions. The over-fertilization of cotton with N can produce plants with excessive vegetative growth without gaining additional yield, in addition to providing a greater potential for loss of the N from the soil-plant system. Uptake and utilization of N by cotton has been evaluated in a number of crop production environments and conditions. Peak periods of uptake and utilization of N by a cotton crop commonly occur near the formation of the first pinhead square (PHS) and again near peak bloom (PB).The greatest potentials for losses of NO3- nitrogen in an irrigated cotton production system occurred with pre-plant applications of fertilizer N and also with those occurring late in the season (after PB). The greater levels of N use efficiency with split applications is recorded. The value of split applications of fertilizer N in-season for optimizing cotton fertilization. Therefore, N fertilizer management recommendations for cotton commonly include the utilization of split applications of fertilizer in-season. N applications recommended between PHS and PB (referred to as the "N application window") in relation to crop condition (fruit retention, vigor, and N fertility status) and previous amounts of fertilizer N applied in-season. Utilizing stage of growth and crop condition in N fertilization is an important application of the crop monitoring systems that are being developed in many cotton producing regions. The accuracy of these crop monitoring systems in relation to stage of growth and management practices such as N fertilization, are improved markedly in many cases with the use of heat unit (HU) systems to predict crop phenology.

Nitrogen effects on physiology and yield

The indeterminate growth habit of cotton (*Gossypium hirsutum* L.) plants makes them very responsive to changes in the environment & management. Both extra nitrogen (N) application and N deficiency affect the growth and yield. Leaf photosynthesis and plant carbon metabolism are the bases of crop growth and yield development. Therefore, investigating the effect of N deficit on carbohydrate content of plant tissues can help us to understand the physiological mechanism of low N influencing cotton yield.

Many earlier studies have shown that N deficit is one of major factors influencing cotton plant growth and yield. Low N stress significantly decreases leaf area development, boll retention and lint yield. The leaves of the low N plants had significantly lower sucrose, higher starch and higher total non structural carbohydrate concentrations as compared to high N treated plants. Insufficient N supply during cotton reproductive growth depressed leaf area, leaf net photosynthetic rate, and leaf chlorophyll content, but increased leaf total nonstructural carbohydrate concentration. The decreased leaf photosynthetic rate from N deficiency might be associated with earlier leaf senescence and slower carbohydrate translocation from leaves to fruits. As a result, fruit abscission of N deficit cotton plants increased and lint yield decreased compared to the plants received sufficient N. Therefore, according to soil nutrient analysis and cotton production recommendations, adequate N during boll development is essential for photosynthesis and yield development.

Nitrogen stress and earliness

Continuous nitrogen stress increased the number of days to appearance of the first flower bud and first flower, had little or no effect on the main-stem node at which the first-fruiting branch appeared, and increased flower bud shedding only in the less determinate early cultivar The total number of fruiting points produced up to first flower appearance was reduced by nitrogen stress in both cultivars. Nitrogen stress from planting to appearance of the first flower bud greatly increased the time taken for flowering to occur at the first 30 fruiting points through delaying first flower appearance in both cultivars and increasing the horizontal flowering interval Nitrogen stress applied later generally had little effect on this earliness index. Nitrogen stress from planting to first flower bud appearance strongly reduced the production rate index. However, nitrogen stress from first flower bud to first flower appearance resulted in a rapid rate of seed cotton production in both cultivars over the control.

PHOSPHORUS

Phosphorus is an important nutrient for plant growth and is required for all stages of cotton growth. It is especially important in the early season period. Phosphorus is a fairly immobile element in the soil, and is not lost rapidly. Phosphorus is removed from the soil each year a crop is grown, therefore it is possible that this nutrient could become deficient and has to be replaced. Soil tests are very important to establish levels of phosphorus in the soil. It is important to sample consistently at the same time each year for accurate comparisons. The pH of the soil has a bearing on the form of phosphate that should be used. MAP (mono ammonium phosphate) is better suited to alkaline soils with a pH greater than 7.0, as MAP is a sightly acidic product. For soils that are acidic (pH less than 7.0), DAP (diammonium phosphate) is a better suited product.

Physiological effects of P

Phosphorus is an essential nutrient for plant growth. It has an important function in cell division so is especially important in young rapidly growing plant tissue. Once P is taken up into the plant, it becomes quite mobile. It will be moved to the youngest areas of the plant. If there is a shortage of P, it will be removed from older leaves and relocated where it is required.

P is involved in energy cycles within the plant, so plants deficient in P will be stunted and low yielding. Early season plant vigour is dependant upon an adequate supply, therefore phosphatic fertiliser should be banded close to the plant line where the growing seedling can have ready access to it. As the P granule begins to react with the soil after application, it moves out from the granule. This movement is limited and is only about 5 cm in total.

Mycorrhizae are fungi that grow in the small feeder roots of cotton and surrounding soil. They colonise the cotton plants's roots and develop a network of fine filaments, thus increasing the root to soil contact. This enhances the uptake of immobile nutrients to the plant, in particular phosphorus. Cotton can grow without mycorrhizae fungi, but plants are stunted, weak and low yielding. 'Long fallow disorder' is a condition directly related to the lack of mycorrhizae in the soil.

Uptake & Removal of P by Cotton

Typical values for uptake of P are 30 kg/ha per year. Therefore, there is a net deficit each year, which would be exaggerated by burning stubble compared with methods of stubble incorporation.

Plant tissue testing may indicate a deficiency during the growing season, although it is probably too late to correct the short-fall in that season. Foliar sprays of phosphorus are not an effective way of delivering P to cotton. The most efficient time to apply phosphorus has been shown to be pre plant. Responses have been recorded in low P soils from side dressing phosphatic fertiliser, but an additional response was noted from preplant applications over the sidedressed P. Even though it is a fairly immobile element, the roots will grow towards the concentrated band of phosphorus. The immobility of P means that it will not be lost from the soil from year to year in the same way as nitrogen, but it is possible that this nutrient may be tied up in form that is unavailable to the plant.

Insufficient phosphorus results in severely stunted plants. Leaves are smaller than normal and very dark green in color. Fruiting and maturity are delayed increasing the vulnerability of the crop to late-season insects (e.g., the boll weevil) and diseases (e.g., verticillium wilt). Lint yield, fiber strength, and micronaire are reduced.

Phosphorus deficiency markedly reduced shoot dry weight (SDW), root dry weight (RDW), whole plant dry weight controlled by plant P contents under P deficient conditions, suggesting an internal regulation in addition to the influence exerted by external P supply. The differences in growth existed among cotton cultivars exposed to same P concentration in the growth medium. Cultivars, which were efficient in both P-acquisition and P-utilization, were efficient accumulators of biomass under adequate as well as deficient level of P supply.

The cotton plant has five main growth stages : germination, emergence, seedling establishment, leaf area and canopy development, flowering and boll development and maturation. An effective way of assessing the development, or progression of physiological events, of a cotton plant is with heat units or growing degree days.

Phosphorus (P) is an important nutrient in cotton production for several reasons. It is essential for vigorous root and shoot growth, promotes early boll development, hastens maturity, helps to overcome the effects of compaction, increases water use efficiency, and is necessary for energy storage and transfer in plants. Phosphorus uptake by cotton closely follows the pattern of dry matter accumulation By first flower, the crop has accumulated about 20% of the total seasonal uptake. After this time, the rate of uptake increases dramatically. By peak bloom, about half of the crop's P has been taken-up, and by first open boll the crop has taken up over 70% of its needs. Failure to maintain the uptake demand during the peak period will result in plant stress, yield loss, and a decline in lint quality. Phosphorus uptake is completed by the time the crop reaches the 50% open boll stage.

Improving the monitoring techniques for evaluating the P-nutritional status of cotton will provide the growers with an additional tool for managing crop nutrition. In recent years, spectral radiometric techniques have been utilized for evaluating the nutritional status of crops. Spectral radiometry is a rapid, non-destructive technique, which if proven suitable can significantly improve the ability to monitor crop nutritional status in-season. One of the commonly used instruments is a Minolta SPAD (soil plant analysis development) meter. The SPAD meter measures the difference in light absorption at wavelengths of 430 and 750 nm.

POTASSIUM

It is well known that high yields require good growing conditions. Knowing the plant nutrient requirements needed to sustain highly productive growth throughout the season will be essential to manage the crop for overall greater productivity. Fertilization of cotton with K is a rather complex issue, because soils vary widely in terms of K-supplying capacity and K fertilizer adsorption. Potassium exists as a constituent of some primary minerals from which many soils were originally formed. It is a part of the interlayer of clay minerals such as hydrous mica, and it may become available due to freezing and thawing or wetting and drying. Potassium dissolved in soil solution is in equilibrium with K attached or bound electrostatically to organic matter and the surface of clay particles. Thus, only a portion of total soil K is soluble, in an exchangeable form, and readily available to plant roots. However, at other times K held in a nonexchangeable form in soil minerals can become exchangeable. When K fertilizer is applied to soil, some fertilizer may be bound or trapped within soil minerals so that part of it is either not available or slowly available to plants.

Optimum cotton (*Gossypium hirsutum* L.) yields are dependent on the availability of an adequate K supply throughout the growing season. One of the prevailing assumptions regarding K deficiency in cotton is that fast-fruiting, earlier maturing genotypes are more susceptible to K deficiency than the more full-season genotypes. Supposedly, the compaction of reproductive growth into a shorter time frame with high yielding, early maturing genotypes intenisifies the K demand and need during the period. The varying genetic makeups of the genotypes complicate the issue and make it difficult to prove cause and effect for maturity differences being responsible for any variation in response among genotypes to K. The differences in root system size or effectiveness could be identified as genotypic differences in response to different K fertility levels. Differences in cotton root surface areas have been linked to differences in K uptake and yield response to K fertility. Okra leaf type cotton lines often are earlier in maturity and produce more flowers than their normal leaf-type near isogenic counterparts. By utilizing okra leaf and normal leaf-type near isogenic pairs, one should be able to narrow the genetic variation and provide a more direct comparison of whether earlier maturity leads to enhanced susceptibility to K deficiencies. The leaf-type genetic variation would still exist, but most of the other genetic variation would be minimized.

Another characteristic commonly associated with cotton grown under low K conditions is that the crop can be harvested earlier. Potassium concentrations in cotton leaves are highly correlated with extractable soil K If it is readily available to the roots. It accumulates in cotton leaves and other plant parts. This trait allows the crop to "bank" a small portion of its total seasonal K requirement during vegetative growth and use these reserves later in the grow-ing season when nutrient requirements are high or uptake by roots cannot keep up with growth needs.

To improve plant tissue analysis as an aid in making fertilizer recommendations or diagnosing plant nutrient status, the relationship between plant nutrient composition and crop yield needs to be better defined. Often, however, the correlation between plant nutrient status and yield is relatively low because many other factors can limit crop yields and unless a crop has adequate nutrients, yield can be limited and much of the other crop production expenses will be wasted. In an attempt to diagnose spatial variability in yields, the crop nutrient status should be determined and deficiencies eliminated.

Fields in which yields are being monitored will have considerable variability, and the reasons for the variability often will not be apparent. Leaf K content is one factor that should be checked as a yield-limiting variable. Soil fertility status is important, but soil test results may not provide all the needed answers because of the interaction of K availability and other factors limiting the plant's ability to take up the nutrient.

When K was withheld from the nutrient solution after first square, the plants continued to grow but at a progressively slower rate. The leaf concentration became progressively lower as K content was diluted by production of dry matter. As K became more limiting, much of it in older leaves was translocated to younger actively growing structures, but even with this reuse, it became limiting within a few days. The old leaves remained green & appeared healthy, but measurements of photosynthesis on individual leaves showed that older leaves with much of the K removed were essentially non-functional. Older leaves expressing deficiency symptoms were produced with inadequate available K. If leaves were produced in an adequate-K environment, they apparently did not develop deficiency symptoms even though most of the K was subsequently translocated out of those mature leaves. This finding illustrates the widely held view that "hidden hunger" can indeed be a production problem. Marginal K concentration in the upper leaves is hardly detectable by visual symptoms. The crop needs to have high K concentrations in its leaves early in the season, because having only sufficient K at that time could result in shortages during boll formation. Later in the season, it is difficult to maintain adequate K in the leaves because of the heavy requirements complicated by limited root growth relative to the size of the total plant. As cotton plants age, the ratio of roots to above-ground parts decreases. This ratio appears to continue after flowering and contributes to the difficulty associated with meeting K uptake needs during the fruiting period. Other studies have found that most nutrients are absorbed by young, recently formed roots. As roots age, they become more coarse, heavily suberized, and lignified. Highly functional roots must continuously grow new absorbing surfaces, but during the boll-producing period, root growth slows and nutrient absorption can-not keep up with the demands.

As leaves on plants developed in low-K media, their appearance was influenced by their K content. Progressively more K-deficient leaves were produced. Early evidence of K deficiency was a downward curling or cupping of the upper leaves; this symptom was followed by mild mottling and eventually a severe interveinal chlorosis. These symptoms appeared in the absence of disease deficiencies of other nutrients; Necrotic areas at the margins of the leaves did not appear until the plants severely deficient cotton leaves without disease have nearly yellow interveinal areas with pale-green veins. The margins are often brown. Diseases did not appear because these plants were isolated, but plants grown on similarly low-K media in the natural environment abscised most of their leaves because of foliar diseases. Potassium deficiency symptoms may be confounded with disease symptoms of various kinds because K-deficient plants have increased susceptibility to infection by microorganisms.

Effects on Physiology and Growth

Potassium is an essential nutrient for many physiological processes, most notably for the transport of the products of photosynthesis within the plant. It does this by regulating the salt balance across cell membranes. Improved transport of photosynthate to active growing points may be expected to promote vegetative growth. Increasing K fertility might promote late-season vegetative growth more in an indeterminate cultivar than in a more determinate cultivar.

The rate of leaf area expansion is reduced by K deficiency. The rate of leaf area expansion increases with increasing leaf K concentration to a maximum that occurs at about 3% K. Although higher concentrations of leaf K were observed in some conditions, there was no additional advantage as far as leaf area expansion rate was concerned. In a production environment, there may be some advantage in early or mid-season accumulation of K in plant tissue since it is mobile and can be utilized later in the season to support additional growth if drought interferes with absorption. Although the total amount of K that can be stored in this way is relatively small, the available K in the plant could provide a buffer during stress periods. Leaf growth rates were 14% lower in plants that had only 1.9% K in the leaves compared to plants fertilized with 3% K in the leaves. Leaf expansion rates declined even more as K concentration decreased and were only 59% as great in plants that had approximately 1% leaf K. The rate of photosynthesis was 7% less in cotton plants that had 1.9% leaf K compared with well-fertilized plants. Plants with 1.9% leaf K looked as healthy as plants containing more K. It has long been assumed that 2% leaf K concentration in plants was sufficient; however, information suggests that this concentration is marginal. Both canopy photosynthesis and leaf growth were lower in plants with 1.9% leaf K compared with plants with higher concentrations of K. The young, fully expanded leaves that had only 1% K had 20% lower canopy photosynthesis rates relative to the well-fertilized plants.

The leaves of cotton seedlings had maximum photosynthesis with approximately 1% K. The rates at which main stem leaves and nodes were added were less in plants with lower K concentrations. Leaves were added to the main stem only 90% as fast in plants whose sampled leaves contained 1.9% K compared with plants with higher concentrations. Cotton leaf area expansion and development, however, are very sensitive to temperature when water and nutrients are not limiting. This small difference in association with lower K concentration would not be detected in most field situations, which is another example of plants' "hidden hunger" being masked by other environmental factors.

In well-fertilized fields that were not irrigated, the K concentration of cotton leaves increased until about 3 weeks after flowering, then declined rapidly. However, irrigated cotton that was well-fertilized did not show a decline in leaf K concentration throughout the fruiting period. In crops exposed to drought, K nutrition during the fruiting period is apparently closely linked to water supply. Potassium uptake is associated with total root length and root growth rates. During the fruiting period, there is severe competition between roots and the increasing boll load for both K and photosynthates. If water supply becomes limited during that period, the uptake of K is insufficient to meet fruit-load needs. This deficit causes a withdrawal of K from leaves and a related slowing of growth processes. In a non-limiting nutrient solution environment, K concentration in the leaves remained stable during growth even to maturity.

When leaf concentrations of potassium are low during dry conditions, interpretation of fertility needs becomes more complicated; it is unclear whether low leaf K is caused by low soil K or drought. The field-grown cotton leaf K concentrations increased with age until 3 weeks after flowering began. At that time, the nutritional requirements of the bolls exceeded the ability of the plants to take up nutrients, causing the leaf K concentration to decrease. Presumably, this phenomenon was caused by a combination of less available water and the fact that there were fewer fine roots to support the above-ground plant. As the boll load increased, there were fewer energy-bearing compounds available to support growth of new roots capable of taking up the nutrients needed.

The relative amount of roots compared with the whole plant weight decreases throughout the season. As K becomes limiting, its concentration in the immature and mature leaves decreases. At the time mineral nutrients are being depleted in the leaves, the need for photosynthetically produced sugar is also reaching a peak for the same reason: energy is needed to support fruit growth. Additional flowers and bolls require both minerals and sugars. As their requirements exceed the supply, less K is available to support growth and functioning of roots. Healthy root cell membranes are effective barriers to nutrient passage, and therefore energy is also required for K uptake. Thus, energy provided by the respiration of sugars is essential for K movement from the soil into the root.

The reproductive parts have a higher priority for available carbon and other nutrient resources. They survive at the expense of roots and other vegetative plant parts. Stem growth becomes incrementally slower. The addition of new leaves slows, and the leaves also become progressively smaller due to the lack of nutrients.This process is a natural maturing of cotton known as "cutout." The cutout occurs earlier when nutrients or water is limiting, causing fewer young bolls to survive. Excessively hot weather also compounds nutrient-deficit problems in cotton, causes bolls to form faster, which causes cutout to occur earlier. Higher temperatures also increase respiration, which requires the plant to consume even more energy. However, excessively hot weather causes young bolls to drop because of high-temperature injury, and the available energy may go into producing vegetative growth.

Increasing K concentration above the amount that resulted in additional growth caused increased concentration of K in the leaves. This phenomenon is called "luxury consumption." When K was withheld from the solution, the concentration in the leaves decreased in relation to the degree and length of time of K starvation. If K was completely removed from the solution, the concentration in the leaves decreased to 30% of the well-fertilized status within 10 days. If the starvation was continued for 29 days, the leaf K concentration decreased to only 17% of the well-fertilized condition. When K was restored to the nutrient media, the plants required about the same time to recover days as they did to deplete K in the starved environment. Thus if plants have a marginal K nutritional status and the growing conditions become worse, an immediate management action such as irrigation is needed. Potassium-deficient plants sometimes die prematurely, because they are more susceptible to diseases and nematodes. Inadequate K also limits growth, which may cause early "cutout" and give the appearance of early maturity. However, plants grown under high-K nutrition continue to produce flowers and set bolls over a longer period, requiring more time to mature all the bolls.

Chlorophyll Meter for Detecting K Deficiency

The lack of "on-the- go" soil sensors for nutrient management remains an important void in precision agriculture. New electronic field diagnostic tools are being developed for use in nutrient management, but there is no substitute for a knowledgeable person to scout fields for crop health. The chlorophyll meter (SPAD-502) was developed by Minolta Company as a tool to manage N status of crops. Several researchers found a strong correlation between the meter's leaf chlorophyll measurements and leaf N content. This instrument detects differences only when leaf K concentrations are below 1%. At that concentration, it is too late to correct the problem. A person with careful observations nutritional problem before the leaves reach such a low concentration.

Potassium Deficiency and Diseases

High yield and quality of cotton requires healthy vigorous plants throughout the season. Several investigators have reported an association of K deficiency symptoms and the incidence of verticillium wilt. The symptoms caused by K deficiency are sometimes mistakenly attributed to the wilt. Potassium deficiency often recognized as bronze-colored leaves with necrosis occurring along the margins without a clear border Conversely, verticillium wilt causes necrotic lesions with well-defined borders and rich brown color between leaf veins. Also, when the stems are split open with a knife, brown staining of the interior xylem indicates verticillium wilt. Broadcasting K fertilizer reduced verticillium wilt symptoms. In enclosed chambers,there were not found any disease symptoms even on severely K-deficient plants. Plants with similar K levels grown out-doors, however, prematurely lost all of their leaves due to foliar disease.

Recent advances in remote spectral imaging of crops should improve our capability for mapping K-deficient areas within fields. This information may be coupled with variable rate fertilizer applications to increase precision in fertilization.

Continuous nitrogen stress increased the number of days to appearance of the first flower bud and first flower, had little or no effect on the main-stem node at which the first-fruiting branch appeared, and increased flower bud shedding only in the less determinate cultivar Nitrogen stress from planting to appearance of the first flower bud greatly increased the time taken for flowering to occur at the first 30 fruiting points through delaying first flower appearance in both cultivars and increasing the horizontal flowering interval Nitrogen stress applied later generally had little effect on this earliness index. Nitrogen stress from planting to first flower bud appearance strongly reduced the production rate index. However, nitrogen stress from first flower bud to first flower appearance resulted in a rapid rate of seed cotton production.

Foliar K and yield

Foliar K fertilization by cotton growers, and in research tests has resulted in inconsistent and largely unpredictable yield responses. The foliar K applications over a wide range of soil and climatic conditions revealed that yield increases were inconsistent with location and between years. K requirement of fast-fruiting and high yielding cultivars late in the growing season exceeds plant uptake. The plant root system activity of these high-yielding cultivars begins to decrease at flowering, which is the beginning of high K demand by the developing boll. Plant growth on a soil of limited K availability coupled with restricted root activity has a dramatic effect on its uptake.

The foliar-applied K source may also affect yield response to foliar fertilization. KNO_3, K_2SO_4, $K_2S_2O_3$, KCl and K_2CO_3 were used as foliar K sources and it was reported that KNO_3 increased lint yields, relative to the other sources. Lint yields were increased by foliar application on sandy clay loam. Modifying foliar K solution chemistry has improved K uptake of cotton. Higher leaf and petiole K concentrations at 1, 3, and 7 d after foliar application of KNO_3 with a surfactant (Penetrator Plus) were reported as compared with a non foliar check or foliar KNO_3 applied without the adjuvant. Foliar K increased second-harvest and total lint yields of cotton. Increases in second harvest lint yields indicate that K availability to the plant was marginal or deficient for boll production in the upper part of the plant. Adding Penetrator Plus - buffered foliar KNO_3 solutions to pH 5.5, compared with a pH of 9.4 for the unbuffered solution suggested that K absorption from foliar applications could be enhanced by modifying solution pH values.

MICRONUTRIENTS

Micronutrients are essential elements that are required in smaller quantities for normal plant nutrition. Their role is equally as effective since deficiencies in micronutrients can cause economically significant crop yield and quality losses. Elements classified as micronutrients include zinc, iron, copper, manganese, boron, molybdenum, and chlorine. Zinc, iron, boron, manganese, and copper can form somewhat insoluble oxides that may not be readily available to plants growing in the calcareous, alkaline soils. Zinc and boron deficiencies can effect earliness by delaying flowering and fruiting of cotton plants. Furthermore, zinc deficiencies can be created by excessive phosphorus fertilization. Zinc has been reported to increase fruit retention of cotton. The role of copper in cotton plant nutrition is less clear.

Preplant micronutrient fertilizer applications did not result in a significant increase in cotton lint yields, compared to unfertilized check plots. Micronutrient fertilizer applications also had no significant effect on lint quality / HVI analyses including length, micronaire, and strength. Although fruit retention at positions one and two were relatively high and little fruit was set beyond the second position.

Table 1: Effect of preplant micronutrient fertilizer applications on upland cotton lint yield. Fertilizers were applied in the sulfate forms.

Micronutrient applied	Lint yield (bales/acre)
None	1.49
Zinc (Zn)	1.99
Boron (B)	1.85
Copper (Cu)	1.38
Manganese (Mn)	1.86
Zn + B + Cu + Mn	1.52

(Source Knowler *et al* 1999).

Physiological effects of micronutrients

Micronutrients are essential for normal growth and development processes of plants because these work as mediators or activators of many enzyme systems. Iron is not constituent of chlorophyll but is essential for its formation. Manganese is related to oxidation reduction balance in plants specially in connection with iron and nitrogen metabolism. Zinc is needed by plants in some of their enzyme systems. Further, under adverse conditions of water stress due to high temperature even with good irrigation system, cotton may need additional supply of micronutrients to realize the yield potential. High soil pH and low level of organic matter further limits the availability of nutrients to cotton crop. The deficiency or biological unavailability of nutrients/ micronutrients adversely affects the growth and development of plants for example, effects of nutrient elements on fruiting efficiency has been divided into two groups with respect to fruiting index. A deficiency of one group of elements (P, K, Ca, Mg, B and Zn) limits fruit production to a greater extent than vegetative growth; whereas the deficiency of a second group of elements (N, S, Mo and Mn) restricts vegetative and fruiting growth to an equal extent. Most of the elements in first group may affect fruiting efficiency because they function in the control of carbohydrate translocation. Application of micronutrients increased the seed cotton yield.

Magnesium

An optimum concentration of magnesium has to be made available to the cotton plants for optimum yield. As nutritional disorder becomes more severe, plant experience a shortage of Mg without showing visual deficiency symptoms called 'hidden hunger'. Magnesium is a component of chlorophyll and is required for photosynthesis. Deficiencies may appear as 'purple red' leaves with green veins.

Phosphorus

Phosphorus performs throughout the growing season starting with promotion of a rapidly developing root system. It also promotes the movement of growth substances within a plant. In deficient conditions very dark green leaves with purplish reddening are produced. The ultimate consequences of phosphorus deficiency is yield reduction. At phosphorus is essential for vigorous root and shoot growth. It promotes early boll development, hasten maturity and increases water use efficiency. It is necessary for energy storage and its transport within plant.

Zinc

In cotton plants, photosynthesis, respiration, carbonic anhydrase and chlorophyll concentration are correlated with Zn nutrition. Photo-synthesis and chlorophyll concentration to be at optimal required a minimum Zn of 13 and 14 mg/gram dry weight. Respiration was not influenced by Zn status. Carbonic anhydrase activity increased curvlinearly as Zn status improved from deficiency to adequacy.

Sulphur

Sulphur is an essential component of two amino acids : Methionine and cysteine. These amino acids are key building blocks needed for protein formation. A shortage of sulphur can trigger inefficient dose of plant nitrogen, since both are essential for protein formation when soil becomes water-logged, temporary symptoms can occur which mimic sulphur deficiency but these disappear as some become aerated. Foliar applications can be beneficial in correcting late season deficiencies.

Boron

Boron has been universally recognized as the most important micronutrient for cotton production. Deficiency can be a problem in soils containing insufficient B, or when availability to the plant is reduced as it changes form with higher soil pH. B may be applied either to the soil before planting or as a foliar application just prior to bloom. Soil- applied B increased cotton yields even when deficiency was not evident in the plants. Foliar-applied supplements and soil-supplied B can correct its concentrations in cotton. Foliar application of B accelerates the translocation of nitrogen compounds, increases protein synthesis and stimulates fruiting. Because small amounts are required, foliar application of B may be more efficient than soil application, especially when deficient conditions are suspected. Multiple foliar sprays of B will ensure an adequate supply during flowering and boll development. Preplant soil application plus foliar sprays of B during the season are recommended for soils testing low in available B. The critical level of B for cotton in most soils ranges from 0.2 to 1.0 ppm, depending on the soil pH, organic matter content and texture. B occurs in the soil as an uncharged molecule (boric acid) and leaches readily. B that is held by the soil is associated primarily with organic matter and is released as the organic matter decomposes. Dry weather can trigger a temporary deficiency as organic matter decomposition slows. Furthermore, dry weather slows root growth and limits B uptake.

Physiological effects of Boron

It aids in the transfer of sugars and nutrients from leaves to fruits, plays an essential role in plant cell formation and increases pollination and seed development. It performs a key function in the growth and fruiting process. While B is essential for all stages of cotton plant growth, an available supply is most important during flowering and boll development. Relatively small amounts of B are required to support the process of growth and development of cotton fibers in the boll. Improved fiber quality (fineness, uniformity and strength) has been reported with B application. Total seasonal B requirement averages about 66-200 g B ha-1 for irrigated cotton. B increases the nitrogen and carbohydrate metabolism and sugar translocation in cotton. Its deficiency in cotton may cause small deformed bolls, shorter fruiting branches, deformed leaves, poor fruit retention and reduced lint yields. It can also affect fiber quality, presumably because of the role of B in cell wall growth.

Boron and yield

There are many reports on the growth and yield responses of cotton to soil or foliar applications of B. Reports of yield response to soil or foliar applications have been inconsistent, with some researchers reporting no yield response to B utilizing non-buffered spray solutions, and others reporting yield responses using B spray solutions buffered to pH 4.0. Work in Arkansas, USA, has shown no yield response to soil or foliar applications of B irrespective of soil N status. Soil applied B increased first harvest lint yields by 9%, 4 foliar applications, each at 0.11 kg B ha-1, resulted in lint yields comparable to soil application of B at 0.56 kg B ha-1, and doubling the B foliar rate did not increase yields, but the B petiole concentration was significantly increased. Lint yield, boll production, flower production, boll retention percentage, and fiber properties were unaffected by soil or foliar applied treatments. However, foliar B fertilization resulted in leaf blade B concentrations of 154 mg kg-1 without detrimental effects.

Calcium

Calcium has recently been discovered to be involved and is thought to limit stress induced production of anti-oxidants. Calcium is an essential component of plant cell walls. Although most plant nutrients can be translocated from older to newer tissues, calcium is considered to have little, if any, translocation in the plant as calcium is bond in plant structures. Newly developing tissues must be supplied calcium via root uptake. Calcium is also known to be 'tied up' in soils with pH above 7.5 by interacting with phosphates to form an insoluble compound. Most low desert soils are above 7.5 in pH and may experience this calcium / phosphate interaction over time unless acid has been added to reduce the soil pH.

Foliar treatments did result in increased cotton value / acre but differences noted were not consistent across field. Increased value noted for fertilizers was associated with areas of field with lowest retention rates in the untreated check at three weeks post application.

These are required in minimum dose or trace levels and have the direct effect on nodules formation, enzymes synthesis and other physiological functions. Generally, yellowing of leaves and shortening of internodes exhibit micronutrient deficiency in plants. This results in reduced crop yield.

Deficiency symptoms

Deficiency in nitrogen results in yellowing and drying of lower leaves, reduction in boll numbers, boll weight, seed and lint index.

The phosphorous deficient plant shows dark green leaves, dwarf in stature. Phosphorus deficiency may delay crop maturity. The fruiting efficiency may also be reduced.

In potassium deficiency situation, the leaves become yellowish with white mottling. Yellowish green and brown specks occur at tips and margins between veins. Leaves finally become reddish brown.

Calcium deficiency

Deficiency in calcium results in excess crop growth with few fruits. It occurs mostly in sandy soils.

Magnesium deficiency

Magnesium deficiency produces purple red leaves with sharply defined green veins. The deficiency also causes reddening of leaves. Foliar spray of 5 per cent Magnesium sulphate with 1 per cent urea dissolved in one litre of water may be sprayed once or twice.

Sulphur deficiency

Sulphur deficiency causes yellowing of young leaves and retards growth. The deficiency can be met by applying sulphates of other nutrients, wherever applicable.

Zinc deficiency

In zinc deficient plants the leaves turn thick and brittle with the margin turned upwards.

It is more pronounced during the monsoon and winter seasons. Zinc deficiency can be corrected with foliar application of 0.5 per cent zinc sulphate sowing.

Iron deficiency

Iron deficiency symptoms are yellowing of leaves, curling and burning on the margin of leaves and stunted growth. It can be corrected by foliar application of 2 per cent famous sulphate.

Boron Deficiency

Boron deficient plants produce die back of terminal buds which do not blossom and drop heavily in production : The young leaves yelol wish green, flower buds are chlorotic.

Copper is rarely deficient in cotton growing environments. In the rare event of copper deficiency, the younger leaves turn yellow.

□□□

9

Photosynthesis

A cotton plant as a biological organism, is the sum total of its biochemical reactions including various life processes of the cotton plant and the biochemical mechanisms that underlie these processes. Often, these processes are adversely affected by certain abiotic and biotic stresses that are regular occurrences in the environment. The negative impact of these stresses is a major factor reducing crop yield.

The process

Cotton plants are photoautotrophic organisms, capable of using light and carbon dioxide for growth. Photosynthesis, the conversion of light energy to chemical energy (photo) and the utilization of the chemical energy (Fig. 9.1) for the formation of carbohydrates from carbon dioxide (synthesis), is a central process in the life of a plant and ultimately determines the overall capacity for growth and reproduction i.e., yield. Photosynthesis is a highly integrated process involving complex interactions between the light and dark reactions, both of which take place in the chloroplasts of leaf cells. In reality, each of these "reactions" represents several complex biochemical pathways, each catalyzed by numerous enzymatic proteins.

The rate-limiting step

The rate-limiting step in the overall process, the fixation of atmospheric carbon dioxide is catalyzed by the enzyme Rubisco. Rubisco is an abbreviation for the enzyme, Ribulose-1,5-bisphosphate carboxylase/ oxygenase. The enzyme is bifunctional, catalyzing the carboxylation of the 5-carbon sugar-phosphate ribulose bisphosphate (RuBP) to form two molecules of 3-phosphoglyceric acid or the oxygenation of RuBP to form one molecule each of 3-phosphoglycerate and 2-phosphoglycolate. The ability of Rubisco to use oxygen as an alternate substrate is costly because oxygen

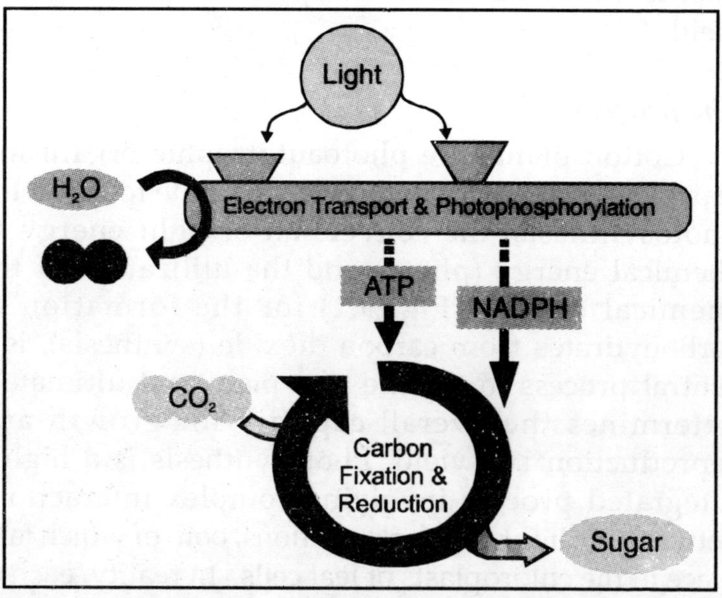

Figure 9.1. *Light reactions of photosynthesis*

competitively inhibits the carboxylation reaction and the fixation of oxygen leads to a net loss of carbon through photorespiration. In addition to these problems, Rubisco has a low affinity for carbon dioxide and a relatively slow rate of catalysis. Together, these properties make Rubisco rate-limiting for photosynthesis under conditions of adequate light.

The regulation of Rubisco imposes a limit on the rate of photosynthesis by affecting the amount of Rubisco available for photosynthesis. Rubisco regulation involves some peculiar properties of the enzyme that passively convert it from an active to an inactive form and a mechanism that actively reconverts it back to the active form. This mechanism involves a second chloroplast enzyme called Rubisco activase. By controlling the switching of Rubisco from an inactive to an active form (called activating the enzyme) activase ultimately determines how much of the Rubisco is in an active form.

The proportion of Rubisco in the active form is often called the "activation state" of the enzyme. Activase is a soluble chloroplast ATPase, a member of the AAA+ superfamily of proteins. Like other members of this family, activase functions as a molecular chaperone, interacting with a target protein, in this case, Rubisco. The chaperoning action of activase facilitates the unfolding of certain loop regions of Rubisco, thereby converting Rubisco from an inactive to an active form (Fig. 9.2). Environmental conditions that affect Rubisco, activase or the interaction between the two enzymes will influence the rate of photosynthesis by changing the proportion of Rubisco that is active. For example, since activase requires ATP and is inhibited by ADP, its activity is adversely affected by a condition like high carbon dioxide that reduces the ratio of ATP to ADP in the chloroplasts. Another condition that negatively impacts photosynthesis by inhibiting Rubisco activation via activase is high temperature.

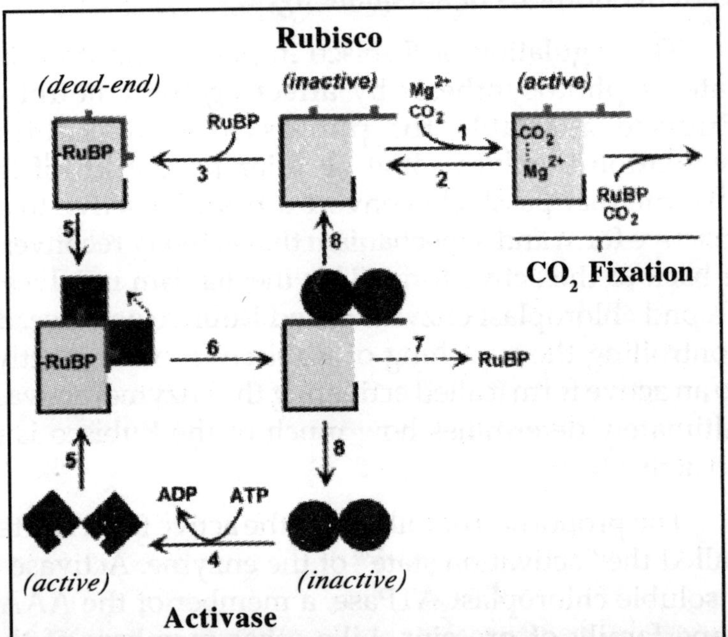

Figure 9.2. *Conversion of Rubisco from an inactive to an active form by activase.*

High temperature reduces photosynthetic reactions

High temperature reduces crop yield by inhibiting photosynthesis. Yield reductions can also occur from an inhibition of pollination. However, unlike photosynthesis, pollination, generally occurs during a relatively narrow portion of the plant's life span. Also, once pollinated the developing fruit is totally dependent on the supply of photosynthate. In cotton, photosynthesis is inhibited when leaf temperature exceed about 32°C. During the summer months, daytime temperatures can exceed 45°C and relatively humidity can be as low as 4%. Because of the high temperatures, cotton cultivation in this arid region is only possible if ample water is available to the crop through irrigation. In well-watered plants, transpiration of water through open stomates evaporatively cools the leaves, reducing their temperature by about 10°C. This cooling is sufficient to prevent inhibition of photosynthesis on all but the hottest days. However, even well watered plants will experience some heat stress on the hottest days, and heat stress will be severe if the hottest days occur at the wrong time, for example, near the end of a watering cycle. In cotton production areas severe heat stress can occur even though air temperatures are generally lower. The reason is that the higher relative humidity in these less arid regions reduces the capacity for evaporative cooling.

The amount of inactive Rubisco increases under heat stress paralleling the loss of photosynthetic activity. This loss of Rubisco activation in response to high temperature occurs before any other plant process is adversely affected. Detailed biochemical studies have shown that the loss of Rubisco activation appears to be related to both an inability of activase to keep pace with a faster rate of Rubisco deactivation and an exceptional sensitivity of activase to thermal denaturation. Understanding and improving the thermal stability of activase may provide a means of increasing the thermal tolerance of plants.

Inhibition of photosynthesis by high temperature

High CO_2 and photosynthetic rate

The effects of elevated levels of carbon dioxide on photosynthesis represent another area of interest these days. Fossil fuel burning is increasing the levels of carbon dioxide in the environment, with consequent effects on the global climate. Because Rubisco has a low affinity for carbon dioxide, photosynthesis should increase as atmospheric levels of carbon dioxide rise, increasing productivity and converting some of the excess carbon dioxide into biomass. However, the actual increase that occurs is generally lower than predicted. Research has shown that the activation state of Rubisco decreases with carbon dioxide due to a reduction in the ratio of ATP/ ADP, which in turn inhibits the ability of activase to keep Rubisco in an active form. Understanding and improving the affinity of activase for ATP may provide a means of increasing photosynthesis and hence plant productivity under elevated levels of carbon dioxide.

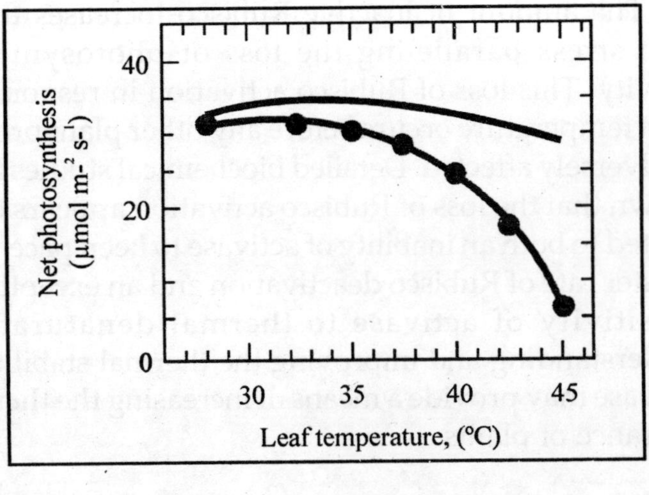

Drought and photosynthetic rate

During drought, photosynthesis is decreased due to stomatal closure, leading to a reduction in carbohydrates, leaf aging and early cutout. Water stress results in various hormonal changes, causing increased shedding and reduced yield. The net effect of water deficit is decreased boll set during the stress, a long delay before boll set fully recovers and premature cutout if the stress is sufficiently severe. The cotton fruit production and retention are important because boll number is highly corrleated with yield at harvest. The boll production and leaf area development are intrinsically linked to leaf area. Because cotton is indeterminate, scientists speculate that the duration of boll formation depends on a balance between photosynthetic capacity and the assimilate demand by growing bolls. The abrupt cessation in boll formation, commonly called "cut-out" in cotton, is thought to result from a shift in balance between the plant's photosynthetic capacity and assimilate demand. Cut-out coincides with a hiatus in vegetative growth and realization of maximum leaf area. Although a direct relationship between cut-out and assimilate production has never been clearly established, most scientists recognize that factors which increase dry matter production also increase boll retention and final boll number.

The imbalance between assimilate production and demand of growing bolls is complicated by the reduction in photosynthetic capacity of leaves after flowering. The photosynthetic capacity of leaves subtending bolls declines after flowering, when the boll demand for assimilates is greatest. Thus concluded that the photosynthetic capacity of the subtending leaf is out of phase with assimilate requirements of the boll. In addition, the bolls in the lower canopy must import about 60% of their carbon from leaves distant to that growing boll.

Water deficits have a major effect on a plant's photosynthetic capacity. Reduction in photosynthetic activity and increases in leaf senescence are well documented and adversely affect cotton production. Other adverse effects of water deficit on the plant's photosynthetic capacity include reduction in cell growth and enlargement, leaf expansion, translocation, and transpiration.

Cotton breeders typically evaluate their advanced strains over diverse environments to identify those genotypes which are most widely adapted to the occurrence of water deficits, and the adverse effect of plant water stress on cotton morphology and physiology. The Multi-Adversity Resistance (MAR) cotton breeding program release after finding that the genotype frequently produced higher yield and had better fiber quality than other short season cultivars under drought e.g. HQ 95 maintained greater leaf area than the other short season cultivars when drought conditions prevailed. These observations led to the hypothesis that HQ95 may be able to maintain greater leaf area, and thus higher photosynthetic capacity under late season water stress than other short season cultivars. Furthermore, the differences observed between HQ95 and other short season cottons under drought may provide an opportunity for us to improve our knowledge of drought resistance in cotton.

Several reviews have outlined the diminished rate of net CO_2 assimilation in water-stressed plants, which can be found for a wide spectrum of species. The decrease of CO_2 assimilation in salinity - exposed plants was not as unequivocally outlined. A direct comparison of water stress and salinity showed that water stress was much more inhibitory to CO_2 assimilation than salinity at an identical Y, which was attributed mainly to the difference in osmotic adjustment under the two stress conditions. Because osmotic adjustment may lead to stress acclimation, this study should be carried out under actual conditions allowing acclimation.

Repeated cycles of water deficit in cotton reduced the sensitivity of photosynthesis to a subsequent water deficit, which was attributed to altered stomatal response to a given Y or to reduced chloroplast response. The capability of chloroplasts to acclimate to water stress was also confirmed by the findings that under decreased Y chloroplast volume was maintained and that the rise in their solute concentration exceeded that of the whole leaf.

The capability of field-grown plants to acclimate to water stress is greater than that of plants grown in small containers, probably because of the lower rate of moisture withdrawal from the root zone. It was in fact shown that the faster the rate of stress development, the more drastic was the decrease in net photosynthesis and leaf conductance. Acclimation obtained at a low rate of Y decrease was associated with solute accumulation, whereas this did not occur at a high rate.

The response of photosynthesis to stress in salinity-acclimated plants is much less documented, and it is quite possible that the more limited inhibition of photosynthesis by salinity is related to such acclimation.

PGRs and Photosynthesis

Chemical hormones produced a marked effect on the photosynthetic activities during the most active period of the life cycle of the plant on account of their greater accumulation in the leaves. More carbohydrate food in the form of hexoses was available for the development of the flowers and bolls when the plants were sprayed with the hormones. As hexoses provided the chief building material, their higher concentration at all stages in the treated plant was a remarkable result. Increase in photosynthetic activity of the leaves also affect the growth made by the plants as the metabolical carbohydrates supply the chief material which builds the plant body. The effect of the treatment by hormones is very much evident when the plants approach the reproductive phase. The dry weight of the plants treated with hormones was higher than those of the control plants. Plants treated with NAA attained the highest weight.

As the total dry weight per plant is the sum of the total dry weight of the leaves, stems and fruiting parts, it is necessary to determine whether the application of hormones affected the growth of the leaves, stems or flowering parts, or any two of those organs or of all the organs. No effect on the growth of the leaves of any one of the chemical hormones was seen till the commencement of the reproductive phase of the plant. As the plant approached the reproductive phase, the enhancing effect of hormones on leaves was visible. This indicates that the leaf growth was favourably affected at the stage when the leaf growth reached its maximum. As the leaf weight is known to be proportional to the leaf area, the total leaf area was higher in the treated plants than the control plants at the time of maximum leaf area stage. The greater the leaf area and consequently greater photosynthetic activity on account of hormone application resulted in better flowering and fruiting activity in the treated plants. The dry weights of the flowering and fruiting parts of the hormone treated plants were higher than the dry weights of the corresponding parts in the untreated plants. The treated plants had greater percentages of the dry weights of the reproductive parts to the total dry matter of the plant than those of the control plants. The main effects of the hormones were on photosynthetic activity resulting in greater production of dry matter; especially the dry matter of the reproductive parts. The efficiency of the cotton plant for reproductive growth was thus greater in hormone treated plants.

Photosynthesis and Boll shedding

The rate of photosynthesis in leaves affects boll shedding due to altered food supply to developing bolls. Plants placed in dim light for three days abscissed nearly all of the young fruits but did not abscise any fruit fifteen days or older. Dim light increased ABA concentration in fruits which were 6 to 11 days old by 29 to 155%, but did not increase ABA concentrations in younger or older fruits.

Cloudy weather and leaf removal can cause shedding of flowers, squares and bolls. The rate of photosynthesis increases when the CO_2 concentration is higher. This increase in photosynthesis caused by CO_2 enrichment of the atmosphere increases the number of blooms and decreases boll shedding. Decrease in photosynthesis or an increase in photorespiration may delay fruiting and decrease fruit retention. Boll shedding increases when nights are warm. Dim light increases ethylene production in plants. The rate of ethylene production by bolls on field grown cotton increases eightfold when the boll load is heavy. Some of the results indicate that a nutritional stress increased ethylene production in young bolls and this ethylene may be a causal factor in boll abscission. Ethylene analysis shows a negative correlation between sugar content of the bolls and the rate of ethylene production.

Photosynthesis and Yield

Photosynthesis is the greatest limiting factor for high yield of cotton and that limited photosynthate causes boll shedding by changing the amount of ethylene and possibly other hormones.

Improving the yield and quality of both upland and Pima cotton is the ultimate goal of the cotton research laboratories. Investigators conduct research on the genetic and environmental factors that limit the number and the growth of cotton bolls produced by a plant. Boll growth relies on a supply of carbohydrate, which is supplied by photosynthesis. Later in the growing season, nutrients for the growing boll, especially nitrogen, come from the breakdown and remobilization of Rubisco and other chloroplast proteins by a physiological process known as senescence. Leaf senescence is still poorly understood at the biochemical level even though this process can limit the supply of carbon for boll growth and negatively impact yield and quality. Understanding the senescence process, especially the loss of carbon dioxide fixation from Rubisco degradation and the effects of environmental stresses such as high temperature on senescence, are topics of interest presently.

Because cotton is grown exclusively in warm regions of the world, it is plagued by insect pests. Whiteflies are sap-sucking insects that derive their nutrition from the sugar-rich phloem tissue of leaves. When the insect is present at high populations, its feeding can reduce yield and deposits of its excreted material known as honeydew damage the cotton fiber by producing what is known as "sticky cotton" (Fig. 9.3). Since whiteflies require plant nutrients for growth, such as sugars (carbohydrates), amino acids, and proteins.

Figure 9.3. *Clean and "Sticky" cotton.*

Photosynthesis is next century: an assessment

Cotton crops grown in future environments will be subjected to a climate for which they were not bred. Several important conclusions regarding the effects of elevated (CO_2), temperature, water, and nutrients on plants can be drawn. The direct physiological effects of higher (CO_2) and photosynthesis and transpiration will have a myriad of secondary effects. More carbon can be fixed in high CO_2 grown plants at all levels of water and nutrient deficient conditions and across a wide range of temperatures. High CO_2 grown plants will have greater stomatal resistance, however, increased green leaf area offsets that effect. Developmental events such as floral initiation, flowering, boll opening, and leaf initiations are relatively intensive to high CO_2. Since cotton is very plastic in its growth, additional carbon available in a high CO_2 environment favoured more vegetative and reproductive growth across a wide range of conditions. This will result in more fruit if temperature conditions are favourable for fruit retention and higher yields.

Cotton producers in the 21st century will be growing crops in different climates than today. Atmospheric CO_2 concentration [CO_2] will likely exceed 700 ml l^{-1} by the latter half of the next century, and the mean global temperature has been predicted to be 1.5 to 5.9°C higher than today. A 5°C increase will quadruple the number of days per year in which the mean temperature is above - optimum for cotton growth, and many days will be so hot that flowers will abscise. Further, the incidence of extreme weather events within a growing season has also been predicted to increase. Agricultural productivity is extremely sensitive to changes projected in the environment and if the projected changes occur, major changes in cotton production will need to take place and both food and fiber will likely be very different from those of today. Doubling the atmospheric [CO_2] results in increased photosynthesis and dry matter accumulation. Season - long exposure to twice ambient [CO_2] (700 ml^{-1}) caused cotton plants growing in optimum temperature to produce 66% more dry matter than plants grown in ambient [CO_2]. Cotton will respond to water and nitrogen deficits to about the same extent in a high CO_2 as to similar stress in today's ambient [CO_2] environment. The projected increases in temperature will cause more important and negative effects on crop production. The flowering process and resultant fruit and grain production is severely limited by temperatures above 30°C. Elevated levels of atmospheric [CO_2], water, and nutrients do not ameliorate the damaging effects of above optimum temperatures on food and fiber production. If the climate change occurs as forecast and greater instability occurs as projected because of increased atmospheric [CO_2] and other radiative "greenhouse" trace gases along with accompanying temperature increases, then our major crops will be subjected to more adverse weather. On the other hand, the vegetative structures of plants are less sensitive to high temperatures. If global warming occurs, man's diet may shift so that more of his nutrition will be derived directly from leaves, stems and roots or animals that consume these structures. Breeding both heat and cold tolerant cultivars would be highly desirable in the temperature climates where seasonal change in temperatures is greater allowing a window of opportunity for adjustments in manipulating cultural practices. For equatorial climates, where seasonal temperature variability are less pronounced and the present-day temperatures are very close to optimum for many summer grown crops, breeding high temperature tolerant cultivars is extremely important if average temperatures increase as predicted. Developing cultivars that can withstand short periods of heat shock and drought stress would be beneficial both in the present-day and future warmer climates.

Developmental processes are very temperature dependent. Average temperatures above 30°C caused young bolls to abscise. Using the database generated from these studies, functional alogrithms were developed and are being incorporated into a physiologically based cotton simulation model, GOSSYM. Such model will allow regional impact and alternative production strategy studies.

To predict the impact of climatic changes on crop production, one needs to model the effects of various weather parameters on crop performance and the interaction of those parameters on crop behaviour. To obtain realistic responses of plants to various environmental factors, it is imperative for plants to be grown at nearly natural levels of solar radiation because of the effect of elevated [CO_2] depends on radiation. Naturally sunlit chambers known as soil-plant-atmosphere-research (SPAR) units that have the capability to control and monitor temperature, water, nutrients and atmospheric [CO_2] systematically can be used in these experiments. Experiments using the SPAR facilities have shown unambiguously that cotton responds to several environmental factors of concern in a changing climate. From these studies, the database necessary to develop response functions for various facets of cotton growth and development has been acquired. The effects of changes projected in the climate on cotton physiology can be incorporated into the simulation model, GOSSYM/COMAX.

□□□

10

Respiration

Respiration is the process of oxidizing food molecules, like glucose, to carbon dioxide and water. The energy released is trapped in the form of ATP for use by all the energy-consuming activities of the cell.The process occurs in two phases: glycolysis, the breakdown of glucose to pyruvic acid and the complete oxidation of pyruvic acid to carbon dioxide and water .The glycolysis occurs in the cytosol. The remaining processes take place in mitochondria.

Mitochondria

These are membrane-enclosed organelles distributed through the cytosol of most eukaryotic cells. Their main function is the conversion of the potential energy of food molecules into ATP. Mitochondria have: an outer membrane that encloses the entire structure and contains many complexes of integral membrane proteins that form channels through which a variety of molecules and ions move in & out of the mitochondrion and an inner membrane that encloses a fluid-filled matrix The inner membrane contains five complexes of integral membrane proteins: NADH dehydrogenase (Complex I), succinate dehydrogenase (Complex II), cytochrome c reductase (Complex III; also known as the cytochrome b-c_1 complex), cytochrome c oxidase (Complex IV) and ATP synthase (Complex V). Between the two is the intermembrane space, the inner membrane which is elaborately folded with shelf like cristae projecting into the matrix and some 5–10 circular molecules of DNA

The matrix contains a complex mixture of soluble enzymes that catalyze the respiration of pyruvic acid and other small organic molecules. Here pyruvic acid is oxidized by NAD^+ producing NADH + H^+ and then decarboxylated producing a molecule of carbon dioxide (CO_2) and a 2-carbon fragment of acetate bound to coenzyme A forming acetyl-CoA.

The Citric Acid Cycle

This 2-carbon fragment (acetyl –co-A) is donated to a molecule of oxaloacetic acid. The resulting molecule of citric acid (which gives its name to the process) undergoes the series of enzymatic steps shown in the diagram (Fig.10.1) The final step regenerates a molecule of oxaloacetic acid and the cycle is ready to turn again. Each of the 3 carbon atoms present in the pyruvate that entered the mitochondrion leaves as a molecule of carbon dioxide (CO_2). At 4 steps, a pair of electrons ($2e^-$) is removed and transferred to NAD^+ reducing it to **NADH** + H^+. At one step, a pair of electrons is removed from succinic acid and reduces FAD to $FADH_2$.The electrons of NADH and $FADH_2$ are transferred to the electron transport chain.

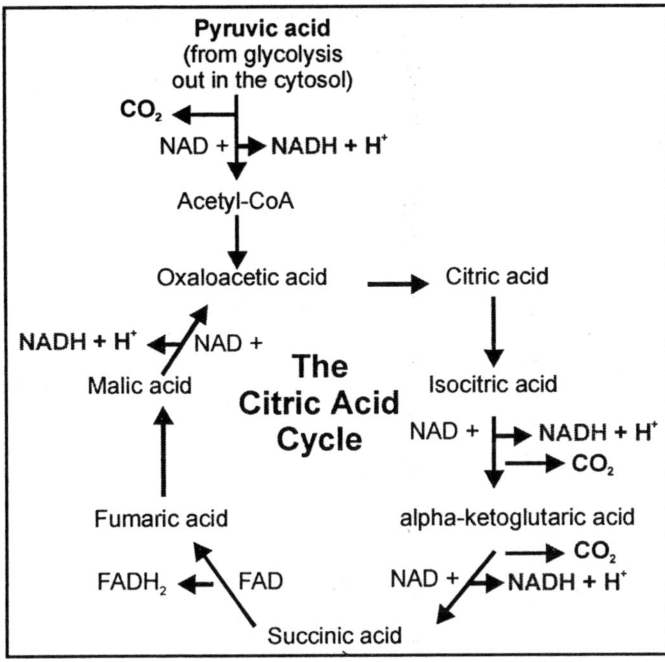

Figure 10.1. The Citric acid cycle

The Electron Transport Chain

The electron transport chain consists of 3 complexes of integral membrane proteins :the NADH dehydrogenase complex (I) the cytochrome c reductase complex (III) and the cytochrome c oxidase complex (IV) and two freely-diffusible molecules ubiquinone and cytochrome c that shuttle electrons from one complex to the next.

The electron transport chain accomplishes:

- the stepwise transfer of electrons from NADH (and $FADH_2$) to oxygen molecules to form (with the aid of protons) water molecules (H_2O); (Fig.10.2)

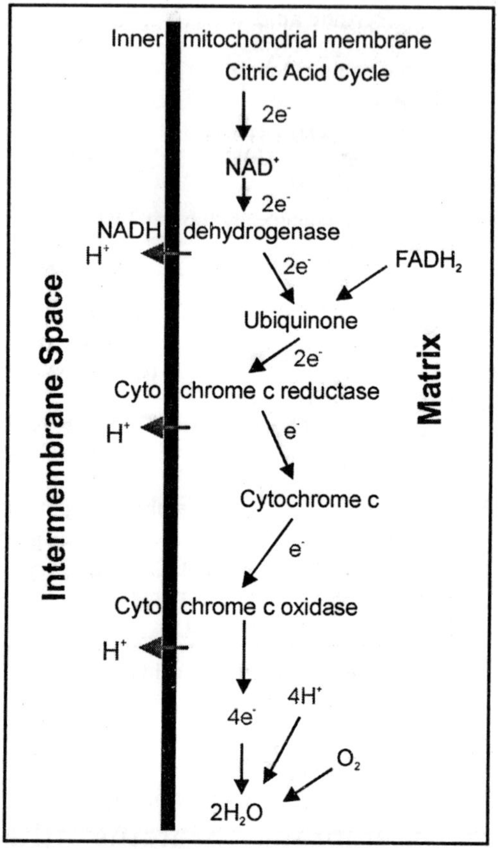

Figure 10.2. The electron transport chain

- harnessing the energy released by this transfer to the pumping of protons (H^+) from the matrix to the intermembrane space.

As Cytochrome c can only transfer one electron at a time, so cytochrome c oxidase must wait until it has accumulated 4 of them before it can react with oxygen. Approximately 20 protons are pumped into the intermembrane space as the 4 electrons needed to reduce oxygen to water pass through the respiratory chain.The gradient of protons formed across the inner membrane by this process of active transport forms a miniature battery. The protons can flow back down this gradient, reentering the matrix, only through another complex of integral proteins in the inner membrane, the ATP synthase complex .

Chemiosmosis in mitochondria

The energy released as electrons pass down the gradient from NADH to oxygen is harnessed by three enzyme complexes of the respiratory chain (I, III, and IV) to pump protons (H^+) against their concentration gradient from the matrix of the mitochondrion into the intermembrane space.

As their concentration increases pH decreases, thus a strong diffusion gradient is set up. The only exit for these protons is through the ATP synthase complex (Fig .10.3) As in chloroplasts, the energy released by these protons flow down their gradient is harnessed to the synthesis of ATP. The process is called chemiosmosis and is an example of facilitated diffusion.

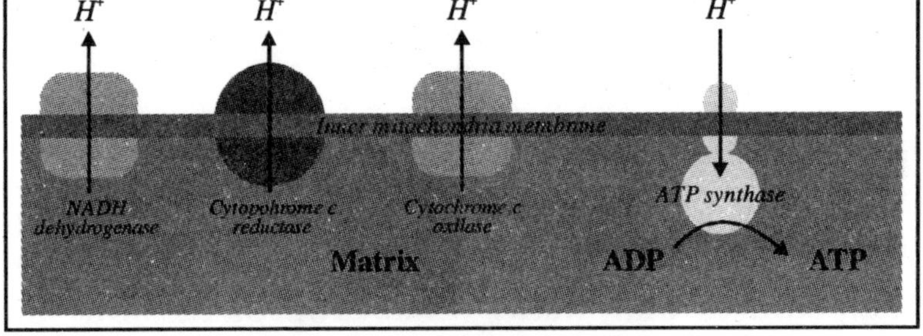

Figure 10.3. *ATP synthase complex*

Most of the ATP is generated by the proton gradient that develops across the inner mitochondrial membrane. The number of protons pumped out as electrons drop from NADH through the respiratory chain to oxygen It is theoretically large enough to generate, as they return through ATP synthase, 3 ATPs per electron pair but only 2 ATPs for each pair donated by $FADH_2$. Thus with 12 pairs of electrons removed from each glucose molecule, 10 by NAD^+ (so 10x3=30); and 2 by $FADH_2$ (so 2x2=4),this could generate 34 ATPs.

Add to this the 4 ATPs that are generated by the 3 exceptions and one arrives at 38.

The three exceptions include

A stoichiometric production of ATP does occur at: one step in the citric acid cycle yielding 2 ATPs for each glucose molecule. This step is the conversion of alpha-ketoglutaric acid to succinic acid ,and at two steps in glycolysis yielding 2 ATPs for each glucose molecule .

But the energy stored in the proton gradient is also used for the active transport of several molecules and ions through the inner mitochondrial membrane into the matrix. NADH is also used as reducing agent for many cellular reactions.So the actual yield of ATP as mitochondria respire varies with conditions. It probably seldom exceeds 30.

Temperature and Respiration rate

Cotton has long been known as a plant which is intolerant of cool weather and thrives at higher tempeatures than many other crop plants. Day air temperatures averaging 30°C and above are common in cotton growing regions and are considered desirable for the growth of cotton. Data shows that some cotton leaves in direct sunlight had temperatures as high as 45.5°C on a day when the air temperature was 41.7°C and the shaded leaf temperature averaged 40°C. Based on this data it appears possible that some cotton leaves may have to withstand temperatures at least four centigrade degrees warmer than the recorded air temperatures at some time during their growth.

Temperature and Respiration Rate:

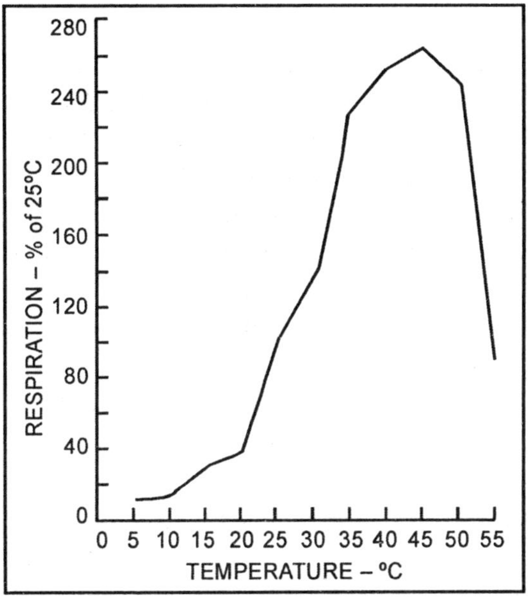

Figure 10.4. Respiration rate at various temperatures

Leaf-discs cut from maturing cotton leaves were used to determine respiration rates at a series of temperatures from 5 to 55°C. It is shown that the rate of respiration increased with temperature upto 45°C and that at 50°C and 55°C there was an initial decrease in the rate of respiration. At temperatures below 25°C there was some tendency for the respiration rate to increase with time over a three-hour period and above 30°C for respiration to decrease with time. At 45 and 50°C the decrease with time was not as great as at temperatures of 35 and 40°C.

In addition to high day temperatures, night temperatures may be even more detrimental to the yield of cotton. Night temperatures have been shown to increase the rates of respiration & reduce carbohydrates Under high night temperatures, more carbohydrates are utilized by the high respiratory rates at the expense of plant growth. In cotton, the loss of carbohydrates and energy caused by high rates of respiration is mainly reflected in fruit abscission, because more carbon from photosynthesis is needed after fertilization for partitioning into developing bolls. The high night temperatures increase respiration, resulting in insufficient carbohydrate produced to satisfy all the plant's needs and also for fiber development.

High night temperatures had no significant effect on respiration after one or two weeks of elevated night temperatures in the field studies. The increase in respiration was significantly higher after the second week of elevated night temperatures. Furthermore, the mean respiration across weeks was also significantly increased under high night temperatures. In the growth room study, respiration in the high night temperature treatments was significantly increased in relation to the low night temperatures.

In the two-week period of high night-temperature , there was an overall numerical trend for a decrease of carbohydrates under high night temperatures, especially of fructose and glucose. However, sucrose showed a significant decrease over time under high night temperatures. The observed trends for the decrease in carbohydrates could be associated with the increased rates of respiration.

The low night temperatures had a more significant effect on the increase of leaf antioxidant enzyme activity than the high night temperatures. Compared to the control, the antioxidant enzyme activity of catalase, ascorbate peroxidase, peroxidase, and glutathione reductase (GR) were significantly increased during the three weeks of the low night-temperature treatments.

There was a consistent trend for elevated night temperatures to decrease the nitrogen (N), potassium (K), magnesium (Mg), and zinc (Zn) concentration of the leaves. Furthermore, the duration of the elevated night-temperature stress significantly decreased the concentrations of K, Mg, and Zn.

Short periods, i.e., one week of high night temperatures, had little or no effect on the night respiration of cotton. However, extended periods of two to four weeks caused significant increases. On the other hand, the photosynthetic activity of cotton was not sensitive to high night temperatures. The decrease of carbohydrates and also the decrease of key nutrients involved in translocation of photosynthates was possibly related to the high rates of respiration, posing a major threat for the carbohydrate supply and flow to the developing bolls. The consistent numerical reductions in fiber weight per seed recorded, together with the significant decrease of fiber weight per seed support the hypothesis that high night temperatures are detrimental for yield of cotton and also cause yield variability due to their effects on respiration and the carbon balance of the plant.

Respiration in water logged cotton

The term 'waterlogged' or 'flooded' is normally used to indicate the unfavorable aeration status of the soil with excessive water levels, leaving little or no room for gasses, especially O_2. This adversely affects plants by curtailing plant growth and development, decreasing absorption of nutrients and water, changing the oxidation state of mineral nutrients resulting in decreased availability or increased toxicity, and by the formation of toxic compounds. The major and immediate effect of waterlogged soils on plant growth is a deficiency of O_2 required for root respiration and growth . This happens because gases diffuse 10,000 times more slowly in water than in air . Hence, soil O_2 supply from the atmosphere is reduced while other toxic gases, such as carbon dioxide, ethylene or methane accumulate to high levels.

Normal irrigation never caused a decline to zero and it recovered quickly, while O_2 in the + waterlogged treatment declined to zero after 48 hrs and by time of the next irrigation had only returned to half of the original level. This 48 hrs time lag might be explained by the fact that even when all the soil pores have filled with water, there are still small but significant quantities of O_2 in the soil that can be used by plant roots or soil micro-organisms for metabolism . If this O_2 is used up prior to the excess water being removed, then root respiration will be jeopardized. In spite of the high clay content of the soil, O_2 content at the surface increased rapidly after irrigation ceased; thus ensuring good drainage of surface water from the field can be expected to have a significant impact in reducing the risk of waterlogging. These results suggest that in this cracking clay soil with 1:1500 slope, waterlogging is not likely to occur unless the soil surface is continuously under water, either from irrigation or rainfall, for at least 48 hrs. The quicker the. excess water is drained, the less chance there is of waterlogging occurring.

The effect of waterlogging on growth and yield depends on the cumulative time that the root system is exposed to an O_2 level less than 0.1%, and/or air-filled porosity less than 0.1 m^{-3}, whether this occurs in a single event or a number of events. The yield of cotton started to decline after 2 cumulative days below this critical oxygenation.

The sum of O_2 reduction between water logged (+WL) and non water logged (-WL) up to each growth stage was significantly correlated to % biomass reduction as shown in Fig. 10.5. Except for the first irrigation in which waterlogging was not achieved, biomass was reduced markedly after each waterlogging event according to the large increase in O_2 reduction. As the soil dried and the O_2 differences between the waterlogged and control plot became less, biomass of +WL plants started to recover, but then fell back again after the next waterlogging event, and remained below the level of the control plants. The linearity of this correlation supports the previous statement that changes in growth & yield are strongly depended on the seasonal cumulative O_2 reduction, whether this occur as a series of short term or one long term waterlogging.

These results illustrate that under waterlogged conditions, the growth and yield of cotton are markedly affected by low soil O_2 status. Photosynthesis seems to be the first plant physiological process that responded to decreasing soil O_2 concentration; prior to the visual symptom of leaf chlorosis. In terms of management, it demonstrates the importance of ensuring adequate bed height and the ability to remove excess water from the field as soon as possible.

Cotton is known to be one of the plants most susceptible to waterlogging due to its low levels of alcoholdehydrogenase (ADH) & pyruvate decarboxylase (PDC), the two main enzymes in anaerobic respiration . The studies have been conducted to provide information for models on the mechanisms involved in the response of cotton to waterlogging. Changes in crop growth rate, photosynthesis, and nutrient uptake especially N & Fe are quantitatively related to the observed differences in soil O_2 and water content, which lead to better prediction of crop losses due to waterlogging. Transgenic cotton lines with enhanced expression of ADH and PDC have been tested for the contribution effect of anaerobic respiration, and the involvement of ethylene by using an inhibitor of ethylene production AVG (aminoethoxyvinylglycine). A better understanding of cotton's response to waterlogging will help in developing improved agronomic management and in breeding to reduce its impact on yield. At present, cotton computer models such as OZCOT contain very simple physiological information about the plant's response to waterlogging.

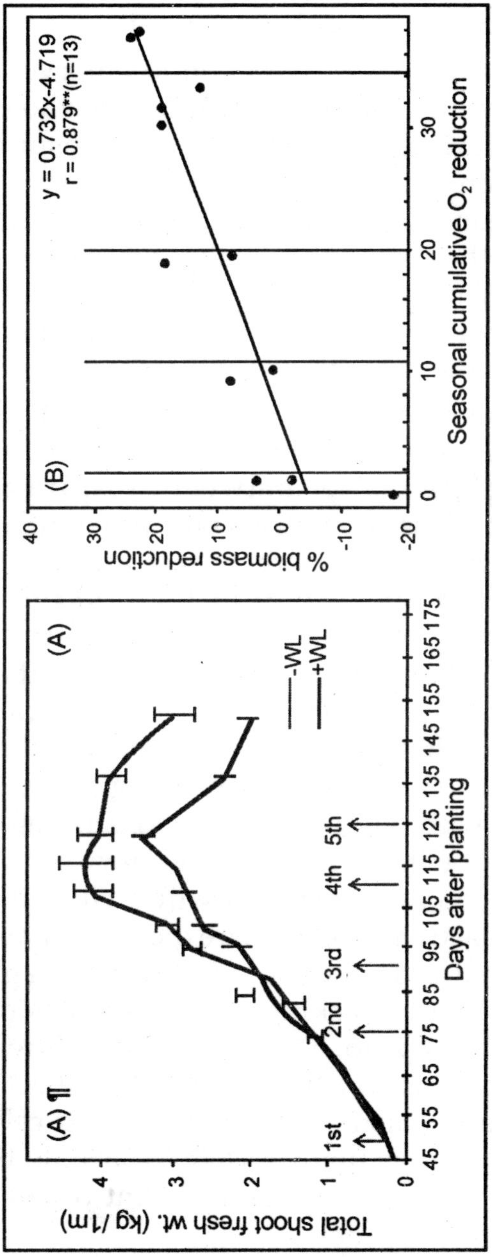

Figure 10.5. Total top fresh weight over the season (A) and correlation between reduction of soil O_2 and biomass compared to control non-waterlogged plots (B). Vertical lines mark irrigation date.(Source Thongbai et al 2001)

Respiration during embryogenesis

O_2 uptake studies with cotton embryos showed that the amount of O_2 consumed by the whole embryo increases linearly with size. If O_2 consumption is expressed on a per cell basis, the pattern is strikingly different than that for the whole embryo. Instead of an increase in the amount of O_2 taken up per cell there is a decrease. Thus, the total O_2 uptake per embryo is correlated with an increase in the total number of cells comprising the embryo.

The histochemical localization of succinic dehydrogenase activity emphasizes an important aspect of the relationship between respiration and embryogeny. That is, with growth, differentiation, and increasing needs for energy, there are differences in respiratory activity within the embryo. This is demonstrated by the differential nature of the localization of enzyme activity. The cells of the embryo in which high enzymatic activity was localized are associated with those areas of the embryo where active growth and differentiation are taking place. This is in agreement with other studies demonstrating the close correlation of increases in respiratory activity with differentiation and active growth. Cells with greater respiratory activity can also be expected to have greater O_2 uptake rates. The result is that cells with greater O_2 uptake rates are masking the lower rates of others. Therefore, except for the relatively undifferentiated embryonic stages in which distribution of activity was found to be equal, the O_2 uptake rates per cell must be viewed as average figures. Only as such do they have value when comparing the different developmental stages of embryogenesis.

The O_2 uptake studies and localization of enzyme activity support the idea that the tricarboxylic acid cycle is operating in all the stages of the embryo studied. Although no experimental evidence is available concerning the immediate stages following the first division of the zygote, aerobic respiration may be expected to occur here as well. This assumption is based upon the observation that the mitochondria have a normal appearance at all stages of embryo-genesis and that the physical location of the developing embryo within the ovule is not one in which O_2 can be expected to be limiting.Data from the inhibition studies lend further support to the idea that the tricarboxylic acid cycle is the only respiratory cycle operating in the stages studied.

Ozone and Root respiration

The reduced allocation of CHO to roots must eventually reduce substrate availability for root growth and maintenance respiration. O_3 reduced substrate availability and the production of root exudates which adversely affected mycorrhizal infection. O_3-induced increases in the translocation of photosynthate to roots have also been observed, particularly with low O_3 concentrations.

Responses of fine root respiration (R_r) to O_3 remain unclear, with studies providing an array of contrasting conclusions.The conflicting observations reflect interspecific variability, contrasting experimental conditions, or inherent variability in measured values of R_r, The studies report chronic O_3 exposures which may have allowed the acclimation of allometry, root system morphology, and physiological function. Such a restoration of homeostasis confounds the interpretation of primary O_3 impacts on R_r.

As R_r may consume over half of the net primary productivity and up to 75% of CHO translocated to roots. The diversity of O_3 effects on R_r noted has hindered the prediction of O_3 impacts on carbon sequestration as part of the overall effects of global change & has prevented appropriate parameterization of long-term O_3 impacts on plant growth and development. O_3 effects on R_r could also serve as early diagnostic signals of O_3 damage to vegetation.

The chronic O_3 exposure significantly increased R_r in cotton while reducing leaf carbon assimilation (A_n). Over short exposures, the O_3 effects on R_r were more variable and hence more difficult to identify,. The acute effects were entirely consistent with chronic impacts. Leaf source strength and root sink activity changed in opposing directions over all time scales considered, with no suggestion of a positive correlation indicative of direct substrate control of R_r. Fine root respiration (R_r) was significantly affected by chronic exposure to O_3. Whereas leaf assimilation of substrate CO_2 (A_n) declined, consumption of substrate CHO in roots (R_r) increased, in cotton.

Thus, Chronic exposure to O_3 reduces carbon assimilation while increasing respiration in fine roots. Acute responses of plants transferred to O_3-containing environments demonstrate that these physiological responses occur within hours, and roughly parallel in leaf and shoot. As these carbon fluxes (A_n and R_r) are in opposing directions, a simple substrate linkage is unlikely. Reduced root growth and O_3-induced reductions in root:shoot biomass ratio are consistent with O_3-limitation of carbohydrate transport to developing roots. The increased respiratory activity may reflect linkage between shoot and root involving demand for nutrients, although this generally results in increased root:shoot ratio. Alternatively, the linkage may reflect changes in the sugar profiles available to support root respiration, possibly attributed to changes in phloem loading in source leaves. Yet another alternative is the transfer of phytotoxic materials from O_3-damaged leaves to roots, inducing damage and enhancing R_r associated with repair mechanisms in roots.

□□□

11

Plant Growth Regulators for Cotton Production

Plant growth regulators (PGRs) encompass a broad category of compounds that promote, inhibit, or otherwise modify plant physiological or morphological processes. Some PGRs are plant hormones or their analogues; others are simply metabolic regulators. PGRs are classified as organic compounds that alter the growth and development of plants. Unlike plant hormones that are endogenously produced by the plant, PGRs may be considered chemical compounds either produced naturally by the plant or synthetically by a chemist. PGRs are biologically active at very low concentrations and elicit responses similar to those observed from plant hormones. Since most plant growth and development processes are regulated by natural plant hormones, these processes may be manipulated by either altering the plant hormone level, or changing the capacity of the plant to respond to its natural hormones.

Modifying cotton growth has become an essential component of cotton production, whether by making adjustments in fertility, water management or use of harvest aids. Applying plant growth regulators to modify early and midseason growth is similar to other management practices. There is potential to influence yield if they are managed properly. The cotton plant is a perennial with an indeterminate growth habit, and reputed to have the most complex growth habit of all major row crops. Furthermore, it is very responsive to management and changes in the environments and responds to any perturbations in its environment with a dynamic growth response that is often unpredictable. Cotton producers and researchers have, therefore, used plant growth regulators (PGRs) and other cultural practices as a means to manage the balance between vegetative and reproductive growth for efficient cotton production.

The key to modifying growth is knowing what that plant needs at each stage of development to reach the final goal of higher yield and quality. The next step is to do everything possible to provide for these needs. Plant growth regulators have the potential to promote crop earliness, square and boll retention, higher nutrient uptake, and keeping vegetative and reproductive growth in harmony to improve lint yield and quality.

Chemical PGRs have been widely used in cotton production in an attempt to adjust plant growth and to improve lint yield and fiber quality. Obviously, the best means to achieve these aims is through the use of desirable genetics in the form of well-adapted high-yielding varieties. However, due to the very varied environment (e.g., soil, temperature, insects, and rainfall) in which cotton is grown, it is almost impossible to have a perfectly adapted variety capable of providing high yields in all circumstances. Therefore, PGRs and other stress management practices have been used in an attempt to consistently achieve high yields.

Natural growth regulators

Several naturally occurring hormones work in the cotton plant to adjust plant growth. When plant growth regulators are applied to the cotton plant, they work in much the same way as the natural regulants already present. In many ways, they supplement or destroy the natural hormone. They often will work together in ratios and concentrations to regulate growth. Some naturally occurring hormones are auxins, gibberellins and cytokinins. Some of the more common plant growth regulators are composed of these hormones or combinations of them.

The auxin family serves a number of functions in plants, such as cell and stem elongation, leaf expansion, increased rooting and changes in fruit abscission. Some yield response has been noted when certain auxins have been applied to cotton, probably due to greater boll retention while other reports have shown no yield response. The gibberellins are associated with stem elongation and some leaf enlargement, but have been shown to increase fruit retention in cotton.

Plant growth regulators (PGRs) play an important role in meeting crop production challenges in cotton as they influence every aspect of growth and development. About 7-35 per cent flower buds and 42-64 per cent of bolls are shed due to physiological disorder and rest due to insect infestation. Thus on an average, cotton plant wastes about 25 per cent of total dry matter due to shedding of buds and bolls. Gibberellins have been found to decrease boll shedding and increase number of bolls. The entire process from anthesis to boll opening takes about 50 days and isolation of plant growth regulators from these developing cotton bolls revealed that contents of auxins, cytokinin, gibberellins were high during initiation and elongation phases. The application of gibberellic acid in field changed leaf orientation for several days and increased plant height. The boll retention capacity was increased with GA_3, IAA and GA_3 + IAA, though GA_3 alone was most effective in improving fruit retention. The application of GA_3 and IAA reduced shedding of buds and bolls; also increased boll weight.

The mixture of GA_3, IAA, IBA and trace elements were not effective in increasing the seed cotton yield in comparison when only cytokinins were supplied. The foliar spray of GA_3 at square formation increased seed cotton yield, however, contradicting reports are also available. Certain PGRs applied exogenously alter the yield parameters markedly. Fibers being one of the important components of yield parameters have attracted considerable attention, thus efforts have been made to improve yield potential of cotton plants in terms of boll retention percent with simultaneous improvement in fiber characteristics such as span length, bundle strength and fineness. From among the various PGRs tried GA_3 + IBA, significantly increased yield due to their effect on plant growth and development. A positive response in lint yield response was observed in GA_3 + IAA treated plants. The hormones cytokinins are produced in the roots of plants. These promote earliness and develop a better root system thus has potential for production of higher cotton yields. The cytokinin treatment at first bloom and two weeks later increased the number of mature bolls on the first and second fruiting position on the sympodial branches, increased total boll set and earliness index. An increase in bud gossypol content with spray of kinetin at match head square stage has been reported.

Cytokinins can promote cell division and enlargement of stems, leaves and fruit. They have also been shown to reduce fruit abscission in cotton, but this is probably an indirect result of making the fruit a better competitor for nutrients. Cytokinins can also delay aging of leaves, allowing longer photosynthetic activity.

The involvement of ethylene as an endogenous regulator in natural leaf abscission has been strongly implicated. Ethylene play multiple roles in leaf abscission resulting in a reduction in auxin transport to the abscission zone and by the induction of hydrolytic enzyme synthesis and enzyme secretion into walls of cell comprising the separation layer. Evidence indicates that ethylene acts as an aging hormone and is related to leaf abscission, at least in part, through its ability to accelerate senescence in these tissues. This view was recently supported by additional findings, and the term "phytogerontological effect" was coined to describe the aging effect of ethylene. Recognizing that the action of ethylene which precedes induction of enzyme synthesis may be achieved by regulation of the supply of hormones to the abscission zone. Ethylene modification of auxin transport is apparently involved in both natural abscission and abscission hastened by exogenous ethylene.

In addition to natural abscission related to senescence, recent reports have implicated ethylene in leaf abscission induced by environmental stress. Ethylene production by freeze-injured leaves was involved in freeze-induced leaf abscission. Increased ethylene production rates by intact cotton petioles during a period of plant water deficit was reported. The level of ethylene production *per se* may not be the controlling factor in leaf abscission, but that the physiological state of each leaf governs the response to increased endogenous ethylene production. The extent of leaf abscission from cotton plants following water stress was related to the severity of the stress in a linear manner.

Synthetic growth regulators

In recent years, synthetic PGRs have been investigated for their ability to alter cotton growth and development in an attempt to control growth and improve productivity. PGRs should be considered as management tools in the producer's arsenal that can be used to ensure efficient production. These compounds are diverse in both their chemistry and use.

In the past two decades, many new PGR compounds have been developed or "accidentally" discovered and subsequently tested on cotton with variable and some-times disappointing results due to a variety of reasons, including poor chemical compounds, changing environments, and varied production practices. Producers have bombarded with a barrage of PGRs and incredible claims about their benefits. Usually there are no creditable data to support these claims on which producers can base their decisions about using these PGRs. This necessitates field studies with normal cotton production practices, as well as growth room evaluation for mechanistic studies. Recent research has focused on the physiological effects and underlying mechanisms of PGRs in an effort to adapt their use to the growth requirement of specific crops and environments. Specific objectives of the research program, depending on the compound concerned are :

- Compare commercially available PGRs in field tests for effect on growth and yield.
- Determine the optimum application rate and timing for the more promising PGRs.
- Investigate the effect of PGRs on emergence, root growth, and seedling development.
- Determine the mode of action of the more promising PGRs.

The information from these studies is used to adapt the technology to best fit into current cotton production systems, and to provide information for farmers to select and use PGRs.

PGRs encompass a broad category of compounds that promote, inhibit, or otherwise modify plant physiological or morphological processes. In the past two decades, many PGR compounds have been developed and tested on cotton. However, the majority of them have not proven satisfactory for one reason or another and have either been discarded or are not used very much. The PGRs that have been used and tested in cotton production in recent years include Atonik, cycocel, Crop+2, Cytokin, Early Harvest, Maxon, PGR-IV, mepiquat chloride, Pix Plus, and PHCA. There are also numerous other so-called PGR compounds that have not been listed because they did not prove to be effective in preliminary testing.

It is important to know and understand for what purpose a particular PGR was designed. For example, a distinction is usually made between growth retardants (e.g., mepiquat chloride, Pix Plus, and cycocel) and growth or yield enhancers (e.g., PGR-IV and Cytokin). However, a growth retardant such as mepiquat chloride, by the nature of its effect on plant growth, may often also elicit other positive growth features such as earlier maturity and a yield enhancement. Main advantage of Pix Plus over mepiquat chloride may lie in the improved partitioning of dry matter by Pix Plus into fruiting structures.

Pix and mepichlor

The cotton plant has a natural mechanism (good boll set) to prevent excessive vegetative growth if nitrogen levels, soil moisture, temperature, insect, disease, nematode controls, and plant populations are all well balanced. In many cases these factors are not well balanced and growth regulators are needed to maintain proper plant size, boll set, and promote early maturity. Additionally, certain growth indeterminate varieties also require plant growth regulator applications to shift cotton from vegetative to reproductive growth. Applying growth regulators based only on stage of development (e.g. pinhead square) is not recommended as plant vigor may vary across fields. Although plant growth regulator applications can be an important part of an overall cotton management program, they can result in reduced yields if applied while plants are undergoing stressful conditions. Before applying plant growth regulators, variety, soil type, fertility, irrigation potential, and field history must be taken into consideration.

Mepiquat chloride is the most commonly used cotton plant growth regulator. It is used to control plant growth and is available under a number of trade names that include Pix, Mepex, Pix Plus, Pix Ultra, and others. Pentia is a new mepiquat product containing a boron molecule (mepiquat pentaborate). This addition of boron is not enough to serve as a replacement for boron fertilization. All mepiquat containing products will be collectively referred to as Pix through the remainder of this section.

Pix (Mepiquat Chloride) is the first plant growth regulator in cotton production to make a significant impact on cotton growth and yield. It has been widely studied and frequently used in the major cotton growing areas. Pix is an anti-gibberellin, meaning that it reduces the production of gibberellin in the plant, which normally would enlarge plant cells. Mepichlor contains the same active ingredient as Pix; therefore it is used in much the same way. Research on the use of Mepichlor is limited.

Pix applications are most successful when applied near the early bloom stage of development and again within 3 to 4 weeks on cotton that is subject to rapid and excessive vegetative growth. Not all cotton fields will benefit from Pix applications. It works best on healthy, non stressed cotton when conditions favour rank growth. Many times, these conditions occur when water through irrigation and rainfall has been abundant, high nitrogen rates have been used, and large, indeterminate varieties have been planted.

The primary effect that Pix has on cotton production is that it takes energy away from leaf and stem; directs it toward boll development and retention. For this reason, it is crucial for cotton growers to understand their individual field conditions and make or omit applications as conditions dictate. As a result of a Pix application at the appropriate times of development, lint is more efficiently produced due to decreased plant height, branch length and leaf size. This provides for a more open canopy that increases boll retention and reduces boll rot; consequently, bolls are given greater potential to produce. However, yield response will depend on the necessity of Pix in a particular field.

Pix management

Several approaches to using Pix have been developed. All have credibility and can be successful in reducing growth, yet a universal approach that is used on every farm has been slow in coming as conditions vary widely throughout the cotton growing areas. The most commonly used approach is to base application decisions on plant height at early bloom. The cotton should be between 20 and 30 inches tall near this growth stage before the first application is made, followed three to four weeks later by a second application if necessary.

One difficulty in implementing the early bloom strategy occurs when the amount of cotton that needs to be sprayed outweighs the amount that the producer can spray in a timely manner. Some producers may prefer to utilize the modified early bloom strategy on a portion of their acreage to manage this situation. The modified early bloom strategy involves the use of height to node ratios and measurement of the most recently expanded internode length. The most recently expanded internode is measured by counting down the plant from the highest mainstem leaf more than quarter size to the fourth leaf. Examine the internode above and below the fourth leaf and measure the larger of the two. This is a good indicator of the plant vigour over that past week or so. Further, growers are advised not to tank-mix with Pix or Mepichlor. Since the active ingredient in these products inhibits gibberellic acid synthesis, their use in combination with PGR IV or Maxxon in a tank mix results in parts of the plant growth regulators working against each other. One promotes vegetative growth while the other slows it. Some researchers suggest their applications be at least a week apart.

Low rate multiple strategy

A drawback to applying these recommended rates in one or two applications is the possibility of drought stress that may set in following an application, creating an adverse effect on the crop. In response to this concern, a low-rate strategy has been developed. This approach uses low-rate multiple (LRM) applications of Pix beginning at squaring and continuing as crop growth dictates. The difference is a more conservative approach to slowing crop growth where applications may be ceased during conditions of stress. If growth is naturally deterred prematurely, growers have less of an investment tied up in the crop than if they used higher rates.

Using LRM (low rate multiple) applications, up to four low-rate applications may be made to the crop. The first application should be made at match head square and subsequent treatments are made 7 to 14 day intervals or when regrowth occurs. Crop growth and growing conditions will dictate application rates and timing. As long as favourable conditions occur, this is an acceptable measure for controlling cotton growth. Care should be taken to scout the crop to ensure that the excessive growth does not require a more aggressive approach.

Plant monitoring approach

Another approach to Pix management is recommended for growers who are willing to employ plant mapping techniques in their management plan. It takes a systematic approach that is based on accumulated plant growth and the current rate of growth. Desired plant height and the assumed number of nodes at the end of the season are key ingredients in this recommendation.

Several physiological assumptions must be made that are applicable to cotton growth in a particular area. The irrigated cotton should produce around 20 to 22 nodes during the season, while nonirrigated cotton may produce 17 to 18. These values may vary from year to year and field to field, yet they are still a good standard on which to base a decision. Growers using this system should establish final plant height goal.

Once a goal has been set for crop growth, a Pix threshold is derived by dividing the desired plant height by the expected number of nodes developed in the growing season. Next, during the first square stages or shortly after, a measurement is taken on plant height and nodes are counted on a representative sample obtained from a field. A grower should take five plants from 10 locations in the field to get an accurate measure. Again divided the plant height at the stage of sampling by the number of nodes. Compare this number with the Pix threshold. If the number obtained during the measurement is less than the Pix threshold, Pix is not required. However, if it is greater than the threshold, further monitoring should be made. Since elongation is most pronounced in the top five nodes, measure the total length of the internodes between these nodes. Divide the inches of this active internode growth by three. A pix application is suggested if the number exceeds the Pix threshold. Pix should be applied at lower rates similar to the LRM approach if a treatment is warranted around match head square. As the plant develops into the early bloom stage and the last application should be made no later than 30 days before harvest.

The approach that growers opt to use depends on the level of management they are willing to expand to reach a goal. The growers should choose the method that they are most comfortable with. All the approaches discussed have credibility and have been successfully used on cotton. The early LRM approach made a greater impact on early season height, node number and internode length. The single early bloom application mostly impacted boll retention. A more effective height control is gained when applications are stated early. However, the effect of application will make on yield is the bottom line.

PGR IV and Maxxon

PGR IV as a plant growth regulator contains gibberellic acid and synthetic indolebutyric acid. The material has shown consistent growth and yield response in some trials. Maxxon is a plant growth regulator similar to PGR IV, but it contains different quantities of gibberellic acid and synthetic indolebutyric acid. It is to be applied on cotton at match head square and first bloom. The effects of PGR IV and Maxxon on cotton has been studied, both plant growth regulators were applied at match head square and at early bloom. Response of cotton to both PGR IV and Maxxon ranged from 10 to 15 per cent. Though not statistically significant, both showed potential for higher lint yields.

The optimum timing for PGR IV on cotton is at planting in the furrow, at the second true leaf stage and at the match head square stage. They found that these applications had the greatest influences on cotton growth, leaf area and square retention. These may be beneficial with an early bloom applications as well. Success may be made with this application if growth is not excessive. To obtain maximum benefit from PGR IV or Maxxon, use only when the crop is healthy and not stressed by heat, drought, insects or poor fertility.

Application of PGR IV has caused earliness, increased leaf area, increased nutrient uptake and square and boll retention as factors leading to higher yields. In furrow treatments have caused slightly earlier emergence and enhanced seedling vigor. A tendency for more bolls to develop on the first two fruiting positions and on the lower 10 reproductive branches following treatment was recorded. Research has shown that at least two-thirds of the total lint produced comes from the first two fruiting positions. PGR IV or Maxxon may be options for growers who want to pursue high yields through a more intensive management approach. It would be advisable to apply the material (PGR IV) early in the growth season to promote early growth, such as in-furrow or match head square treatments, then use Pix or Mepichlor after bloom to slow vegetative growth and force a timely cutout. It is advised not to tank-mix with Pix or Mepichlor. Since the active ingredient in these products inhibits gibberellic acid synthesis, their use in combination with PGR IV or Maxxon in a tank mix results in parts of the plant growth regulators working against each other. One promotes vegetative growth while the other slows it.

Cytokin

The primary component of Cytokin is cytokinin. A yield advantage over the control treatment with the use of Cytokin applied at match head square and at early bloom was shown. When applied in a tank mix with Pix, a greater lint increase was observed. This synergism was not noted with the combination of Pix and PGR IV or Maxxon. The hormone cytokinin is produced in the roots of plants. Under periods of plant stress, especially stress affecting the roots, cytokinin is likely to be inhibited. Cytokin is believed to replenish the plant's cytokinin supply during these periods.

Unlike the other plant growth regulators mentioned, Cytokin may be most beneficial under stressed conditions so maintain proper boll set and retention, promote earliness and develop a better root system. Cytokin or combinations of Cytokin and Pix have the potential for higher cotton yields, though growers are encouraged to use most plant growth regulators judiciously and on a limited basis at first, testing them under their own field conditions.

Salicylic Acid

Finding environmentally safe methods for controlling crop pests or for increasing crop yield is a critical goal for physiologists research. Recent research has shown that salicylic acid may induce several positive plant responses that could help physiologists achieve these goals. Salicylic acid, an endogenous plant substance that can also be applied externally, helps to regulate several plant functions including systemic acquired resistance to pathogens and the formation of flowers. However, there is little data indicating whether salicylic acid applications in the field are effective in increasing crop yield. The effects of foliar-applied salicylic acid on cotton flowering, boll retention, and yield in the field has been tested. Foliar applications of salicylic acid were made either once per year or several times during the season. Because salicylic acid is thought to help diminish some of the negative impact of insect pests, two of the tests were conducted without the use of insecticides. The results showed that salicylic acid had no effect on flower production, boll retention, and yield in the field where no insecticide was used. Although the results were not positive in control, the data should not rule out the possibility that salicylic acid might be an effective treatment in future tests.

The discovery of most commercial PGRs appears to have resulted from seren-dipitous events, and therefore, the nature and action of the chemicals has been quite varied. It is difficult to predict what new variations of PGR will arise. Current trends in the selection and design of specific PGRs has involved multi-entity PGRs, polyhydroxy carboxylic acids, and the addition of micronutrients. It is obvious that the discovery and acceptance of three new classes of PGRs - namely jasmonates, salicylic acid, and brassinosteroids - will lead to the development of new commercial PGRs. The use of future PGRs will probably focus more on the fruit development period and on remedial applications, as well as a more coordinated systems approach to management coupled with more precise crop monitoring techniques.

Chaperone™

The plant growth regulator Chaperone™ has been reported to increase plant nitrogen levels, promote protein constituent transport, and increase overall yields. Chaperone is a combination of nitrophenols, namely sodium 5-nitroguaiacolate, sodium o-nitrophenolate and sodium p-nitrophenolate. Phenolics play a central role in plant metabolism e.g. increased photosynthetic electron transport, improved membrane integrity, increased enzyme / protein production, increased lignin bio-synthesis, and increased fruit retention and growth.

Observances in trangsenic cottons have shown that endotoxin levels have occasionally failed to be fully expressed under various conditions, including environmental factors and varietal differences, thus occasionally leading to less efficient insect control and subsequent yield losses. Cotton plants engineered to express the endotoxin protein, Cry 1Ac, from *Bacillus thurengiensis* (Bt) have shown significant declines in efficacy against *Helicoverpa* spp. during the season, particularly from flowering onwards. There was less total protein in the leaves of older plants as a result of a three-to-five fold reduction in protein synthesis over the season. Thus it was, concluded that not only is there less Bt protein in older plants, it appears that the protein is either less available or less toxic to neonates. The concentration of Cry 1 Ac protein, as a proportion of total protein, also declines during the season. The phenolic properties of Chaperone may aid in transgenic cotton by alleviating non-expression or under expression of Cry1Ac or a combination of Cry1Ac with Cry2Ab, the genes utilized for expression of the endotoxin protein, *B. thuringiensis.*

Data revealed that foliar applications of Chaperone may be a viable means for enhancing lint yields in cotton through the enhancement of plant protein levels. Also, in transgenic (Bt) cultivars the enhanced protein status contributes to improve late-season endotoxin levels, particularly in the squares, resulting in increased mortality of neonate bollworms feeding on the treated plants.

Foliar applied Chaperone under elevated temperatures appears to be a viable means for increasing protein concentration and, therefore, the efficiency of endotoxin expression. Increased protein and endotoxin expression from Chaperone applications under environmental stress conditions is the probable reasons for the increased bollworm mortality and increased lint yields. In addition, improved efficiency from foliar applications of Chaperone may also improve the overall physiological functioning of the cotton plant.

Variability of endotoxin expression and / or concentration within transgenic cotton (Bt) varieties has been and continues to be a concern of cotton growers, researchers, and breeders. While not a consistent problem, it can cause major economic problems. On occasions, transgenic cotton crops require additional late season insecticide sprays due to the probable lack of endotoxin expressed in the cotton plant. This variation in endotoxin efficiency could be due to the inactivation of the introduced gene leading to a reduction in the amount of toxin produced. It is speculated that this lack of endotoxin production may be due to increased environmental stresses such as elevated temperatures and high light intensity. Information on the impact of environmental stresses (particularly elevated temperatures) on protein and endotoxin production is needed in order to make management decisions surrounding current transgenic cotton lines and for the development of new ones. From a management perspective, applications of Chaperone during elevated temperatures might be a means of improving protein and endotoxin concentrations and thereby enhancing yields from increased mortality of bollworms.

Cry1Ac with Cry2Ab, are the genes currently utilized for expression of the endotoxin protein in *Bacillus thuringiensis*. It has been observed that a reduction in the amount of expressed endotoxin occurs as plants mature, leading to a loss of efficacy in the latter stages of the growing season thus increasing the probability of surviving pests which may develop immunity to endotoxin.

Rate and timing of PGRs application

The complex growth pattern of cotton and its dynamic response to management and environment necessitates precise application timing and rate of foliar-applied PGRs for maximum control of plant growth for optimum yield response. Information concerning rate and timing needs to be continually updated as we learn more about each specific PGR (benefits and mode of action) so as to best adapt the technology to our existing cropping systems.

Cotton is often planted under unfavourable planting conditions. Therefore, producers have been interested in PGRs or fertilizer additives for enhancement of seedling growth as well as increased yield. Earlier growth room studies showed enhanced root growth and seedling vigor from using in-furrow seed treatment with PGR-IV and with ASSET. However, field studies have been less than conclusive, often with variable results.

Physiological effects of PGRs and mode of action

A second level of research involves studying the more promising PGRs for their effect on the physiology of the cotton plant and their mode of action. The effects of selected PGRs on photosynthesis, carbohydrate status, membrane integrity, and nutrient uptake have been documented and the results used to explain the mode of action. The use of PGRs to enhance nutrient uptake has also been shown.

Pix Plus has been shown to increase the levels of the polyolmyo-inositol, which is believed to be related to the improved partitioning of dry matter into cotton bolls. Phenolic acids play a possible role as modulators of hormonal activity. The temporal patterns of phenolic acids in cotton fruit in relation to abscission, sensitivity to environmental stress (shade and water stress), ethylene evolution, and abscisic acid concentration of fruits have been reported. This research suggested that phenolic acids modify growth and development of cotton fruit during stress, and indicated the potential for use of phenolic acids as growth regulators in cotton.

Since leaf carbohydrates represent the primary metabolic carbohydrate pools for cotton plants, an understanding of their dynamics during cotton growth and boll development is essential. To understand the dynamics of carbohydrate changes during leaf, canopy, and fruit development and the possibility of influencing carbohydrate translocation out of the leaf using PGRs, carbon labeling techniques were used to show the influence of these on translocation. Certain PGRs, e.g., PGR-IV and possibly mepiquat chloride, can enhance translocation of carbohydrates out of the leaf, which was associated with an increase in leaf photosynthesis and a yield advantage. Photosynthesis is often improved when PGRs are used, but this is thought to be an artifact of the improved movement of carbohydrates out of the leaf, allowing photosynthesis to occur at a maximum rate.

Plant height

Plant height was significantly and consistently decreased by mepiquat chloride and Pix Plus by 15-20%, sometimes with a slight accompanying decrease in main-stem node number. Cycocel decreased height significantly but not as much as Pix; however, cycocel (which is still used on cotton in many developing countries) significantly decreased yield in other related studies. Most other PGRs had either no effect on plant height and node number or a small increase in height due to the growth-promoting properties of the compound. In general, PGRs increased the concentration of certain nutrients in the petiole, but as with yield, the results have been inconsistent. There have been no significant effects of the PGRs tested on fiber quality.

Effect of PGRs on root growth

Poor root growth early has been a serious limitation to optimum yields. PGRs, in addition to fungicide and insecticides, may offer an opportunity for enhancing early-season plant development. The research has shown that in-furrow or seed treatment with certain PGRs is beneficial to early root growth and seedling development in cotton. For example, growth room studies revealed that the in-furrow applications of PGR-IV resulted in dramatic increases in root length, root dry weight, number of lateral roots per plant, and nutrient uptake 1 week after planting. These differences were still apparent 5 weeks later at pinhead square but to a lesser degree. A range of PGR substances have been evaluated in growth chamber studies for their effect on root growth and seedling development. A number of compounds such as PGR-IV and ASSET applied in-furrow or as a seed treatment have shown promise for enhancement of root growth and early shoot development, although field results have not always been consistent. Indole butyric acid (IBA) and Pix plus IBA stimulated root growth, but Pix alone did not stimulate root.

Effects of PGRs on stressed cotton

PGR-IV can have a remedial effect on stressed cotton. For example, under conditions of mild water deficit, PGR-IV can partially alleviate the detrimental effect of water stress, i.e., on photosynthesis. It has also shown similar advantageous alleviation of stress effects under conditions of mild flooding and canopy shading. PGR-IV has also been reported to help under conditions of low temperature. This remedial action of PGRs may ultimately turn out to be one of their more important roles in cotton production in the future.

PGRs and early maturity

The nature of the action of PGRs, e.g., decreasing vegetative growth, increasing fruit set, and improving partitioning to fruit, often causes earlier maturity of the developing boll load. Only Pix has shown a consistent trend toward hastening maturity, as indicated by reaching physiological cutout of NAWF=5 about 7 days earlier, with a slightly larger percentage harvest at first pick. There was not a clear trend for any other PGR compared to the control toward early cutout. In general, mepiquat chloride causes a significantly earlier maturity about 50% of the time.

Field and growth room studies are being conducted to understand the role of *Bacillus cereus,* the new bacterial component of Pix Plus. Information to data implicates *B. cereused* an increased dry matter partitioning to the fruit through improved efficiency of biochemical pathways in the leaf, possibly from bacterial metabolites on the leaf surface.

PGRs and yield

A routine field test comparing commercially available PGRs has been conducted. The rates and timings of the PGRs tested were according to best-scenario recommendations. The PGRs tested include Atonik, Cycocel (CCC), Maxon, Early Harvest, PGR-IV, PHCA, Cytokin, Crop+2, Mepiquat Chloride (Pix™), and Pix Plus (formerly MepPlus). Foliar spray treatment applications were made with a CO_2 backpack sprayer calibrated to supply 10 gal/acre of solution. Fertilizer, weed control, and insect measures were added and experiments were furrow-irrigated as needed throughout the growing season. Routine records were made of plant height and main-stem node number, petiole nutrient concentration, maturity, boll weight, boll number, lint percentage, fiber quality, and yield. The yield response with added PGRs has been variable and inconsistent. There were no significant treatment effects on yield, possibly due to unusually stressful seasons masking any possible PGR effects. Part of the reason for variable yield responses from PGRs is the extremely varied environments and crop conditions under which the PGRs are used, and also the lack of understanding of the nature and performance of the specific chemical compounds. Another obvious reason is that some of these chemicals do not really do the job they are supposed to.

The most consistent response has been shown with PGR-IV (a significant yield increase 43% of the time), compared to PHCA (40%), Cytokin (33%), Pix (29%), and Crop+2 (20%). Early harvest has not shown any significant effect on yield. The average yield increase from PGR-IV was more as compared with Pix. However, Pix is a growth retardant and as such was not conceived as a yield enhancer. Studies have shown significant yield increases from Pix about 25% of the time associated with the controlled plant growth. The effect of PGRs on boll weight and boll numbers has shown no clear trend. Another way to use these data is to consider the percentage increase for each PGR compared to the untreated control. For example, the average response to mepiquat chloride was 2.7%, for PGR-IV it was 4.0%, and for Early Harvest it was 2.0%.

PGRs allow for manipulation of physiological processes in plant growth and development for more efficient crop management and increased yields. The use of PGRs in cotton is a useful production practice for controlling plant growth and enhancing yield. However, the effect of available PGRs on growth and yield has generally been inconsistent & disappointing. This is partially attributed to seasonal variations in weather and crop conditions, but also to poor PGR materials and a lack of previous testing of the chemicals concerned. In some cases, the chemicals are not suited for use in cotton. Good management decisions are therefore necessary in deciding whether to use a PGR and also in selecting the most appropriate compound. Continue research at both applied and basic levels will elucidate the specific effects and mode of action of PGRs and thereby aid in adapting their use to current cotton production systems while also improving their performance and consistency.

□□□

12

Metabolism of Fiber Development

Morphological fiber initiation coincides with flower anthesis. The timing of fiber initiation varies from many hours to several days before anthesis. It appears that developmental cues for fiber production are present over a long period of time, so, there is longer 'window of opportunity' over which development can be manipulated to increase fiber production. Genetic techniques have expanded our understanding of the events leading to fiber development and have identified a pre-anthesis stage where epidermal cells undergo changes in gene and protein expression in preparation for fiber expansion. Epidermal cells on pre-anthesis ovules are 'primed' for fiber development and awaits hormone signals associated with anthesis.

Fiber initiation is influenced by physiological changes in the plant. Number of fibers produced per ovule is affected by planting date and changes in mean minimum daily temperature. Analysis of fiber development *in vitro* indicates that unfertilized ovules require the addition of indole-3-acetic acid and gibberellic $acid_3$ for ovule growth and fiber development. If ovules are fertilized before culturing, then exogenous IAA has a minimal affect on fiber development, but exogenous GA_3 is required for ovule growth. Cytokinins and abscisic acid have an inhibitory affect on fiber development. Analysis of hormone content in young ovules and fibers grown *in plantae* indicates that auxin levels are high initially i.e. at the time of anthesis and then drop dramatically at eight days post-anthesis indicating that an increase in auxin levels is important for epidermal cells to differentiate into fibers.

The mature cotton fiber used in textile processing is the end product of a series of developmental events in the life of this ovule epidermal cell. Thus, improvement in cotton fiber properties for textiles depend on changes in the growth and development of fiber. Fiber perimeter is established early in the fiber developent and remain constant. Manipulation of fiber perimeter has a potential to impact the length, micronaire and strength of cotton fibers. The perimeter of fiber is regulated by biological mechanisms that control the expansion characteristic of the cell wall and establish cell diameter.

The diameter is established at fiber initiation and is maintained throughout the duration of fiber development.Improvement in fiber quality can take many different forms. Changes in length, strength, uniformity and fineness are all needed to accommodate new techniques in textile spinning. Fiber perimeter is the variable that has the greatest effect on fiber elongation and strength properties. The mature dead fibers have an elliptical morphology,whereas living fibers have a cylindrical morphology during growth and development. Geometrically, perimeter is directly determined by diameter. For this reason, understanding the biological mechanism that regulate fiber diameter is important for long-term improvement in cotton.

Growth Phases

Cotton fiber progresses through four developmental stages : initiation, elongation, secondary wall synthesis, and maturation. Fiber initiation involves the initial isodiametric expansion of the epidermal cell above the surface of the ovule. This stage may last only a day or so for each fiber. Because there are several waves of fiber initiation across the surface of the ovule, one may find fiber initials at any time during the first five or six days post anthesis. Fiber initiation may occur in two morphological phases, a spherical expansion above the epidermis (ballooning) followed by elongation in which the rounded cells develop a tapered end. Not all the cells of the epidermis develop into fibers. Between 10 per cent to 25 per cent of epidermal cells become fibers.The elongation phase encompasses the major expansion growth phase of the fiber. Depending on genotype, this stage may last for several weeks post-anthesis. During this stage of development the fiber deposits a thin, expandable primary cell wall composed of a variety of carbohydrate polymers. The final length of fiber is the product of the rate of elongation per day and the total period of elongation which is a genetic attribute. Variability exists among different cotton varieties in the rate and the period of elongation and secondary wall deposition. Earlier concept was that the secondary thickening phase does not begin unless the elongation phase is completed. Later definitive work has shown a considerable overlap between the elongation and secondary thickening phase. Through genetic manipulation it may be possible to alter either elongation or dry weight increase of the fibers without appreciably changing the other. Thus fiber elongation and secondary wall deposition are not necessarily controlled by the same genetic factors. Secondary wall cellulose deposition in fibers begins in advance of cessation of elongation at a time related closely to final fiber length. The cause and effect relationships are not established but it is possible that onset of secondary wall elongation controls fiber length even though elongation continues beyond the point of beginning of secondary wall thickening. The deposition of secondary wall does not immediately stop fiber elongation as if specific wall thickness is required to stop elongation completely. Elongation occurs throughout the length of fiber and not just at the tip, though more rapid at the tip region. The fibers at early stage of secondary wall deposition elongate solely by the tip growth or the cellulose microfibrils are deposited in a random or transverse orientation until distinct increase in birefringence typical of secondary wall appears. As the fiber approaches the end of elongation, the major phase of secondary wall synthesis starts. The secondary wall is composed almost exclusively of cellulose. This stage lasts until the boll opens (50 to 60 days post-anthesis), the cell wall now becomes progressively thicker and the living protoplast decreases in volume.

The establishment of fiber diameter is a complex process that is governed to a certain extent, by the overall mechanism by which fibers expand. The expansion of fiber cells is governed by the same related mechanisms occurring in other walled plant cells. Most cells exhibit diffuse cell growth, in which new wall and membrane materials are added throughout the surface area of the cell. Specialized, highly elongated cells, such as cotton fibre, root hairs and pollen tubes, expand via diffuse growth. Later in the fiber development when wall starts secondary cell wall synthesis, the organization of cellular organelles is consistent with continued diffuse growth. Many cells that expand via diffuse growth exhibit increase in both cell length and diameter, but cells that exhibit tip synthesis do not exhibit increases in cell diameter.

Cell expansion is also regulated by the extensibility of the cell wall. Regardless of whether cell expansion occurs via tip synthesis or diffuse growth, the wall in the region of expansion must yield to turger pressure if the cell is to increase in size. For this reason, cell expansion most commonly occurs in cells that have only a primary cell wall. Primary cell walls contain low level of cellulose. Production of the more rigid secondary cell wall usually signals the cessation of cell expansion. Mature fibers exhibit thickened secondary walls composed of about 94 per cent cellulose and spiral twisting. The degree of thickening and the angle of spirals affect fiber strength. At the time of fiber maturity, the ovary wall splits and opens along loculur suture lines, leaving seeds and fibers exposed. This leads to loss of water and collapse of fiber cells. The lumen contents dry up to form a residue.

Wall synthesis can be monitored through the measurement of changes in form birefringence.The birefringence is a measure of the amount of ordered material in an object, for example, the amount of ordered rods in a less refractive medium. Cell walls have form birefringence due to the crystallinity of the cellulose microfibrils and their organization in a matrix of non-cellulosic material. As the amount of ordered cellulose in the fiber walls increase the birefringence increase. The greater the form birefringence and brighter the fiber appears. Prior to twenty days post-anthesis fibers exhibit no detectable wall birefringence. Between 20 and 40 days post-anthesis fibers exhibit a dramatic increase in birefringence indicating deposition of highly ordered arrays of cellulose microfibrils in the secondary cell wall.

The observed increases in wall birefringence coincide with higher rates of wall synthesis and dry weight accumulation in fibers. The increase in dry weight is due to synthesis of cellulose and wall developed is composed almost exclusively of high degree of polymerization cellulose. The secondary cell wall contains little non-cellulosic component and no lignin. The secondary cell wall must be plastic enough to allow for the observed increases in fiber diameter. The secondary cell wall is flexible enough to allow for the observed increase in size. No significant increase in diameter is reported between 30 and 40 days post-anthesis in some species, however, *G. barbadense* did exhibit significant increase in diameter during this period.

Fiber perimeter is an important textile trait that directly affects quality properties, such as fiber length, strength and micronaire. The literature is contradictory regarding whether or not fiber diameter changes during development. Fiber perimeter is thought to be established early in fiber development and remain constant. It is regulated by biological mechanisms that control the expansion characteristic of the cell wall and establish cell diameter. Fiber cell regulates the relationship between increases in length and diameter as in both cases, the cell wall must expand or stretch. However, the direction of expansion in each case is different. The patterning of microfibrils in the cell wall has been proposed to regulate wall expansion characteristics. Modulation of cellulose microfibril patterns in the developing cotton fiber may be a viable method for changing cell wall expansion properties and thus, shifting expansion away from increases in diameter and toward further increases in length. Fiber diameter is thus, a dynamic property that changes during growth. If diameter change can be manipulated, then improvements in textile trait of fiber such as fineness can be achieved.

Cytological features

Developing cotton seed is covered by epidermal layer which is composed of epidermal cells, and cotton fiber initials. The developing fibers receive nutrition from pigmented layer of seed-coat underlying the epidermis, which is several cells thick and is supplied with vascular tissue. The basal part (foot) of the fiber is absorptive in function. The epidermal cells are closely packed, cuboidal and rich in cytoplasm containing a large nucleus. The morphological differentiation of fiber begins when an epidermal cell rounds up and protrudes. The external surface stretches outward, and the cell balloons above epidermal surface. Once elongation has begins, the fiber cells do not divide. The transition of elongation phase starts slowly as blunt tipped cells elongate towards micropylar end. At second and third day following anthesis, the rate of elongation of fiber increases which exceeds the rate of diametric expansion so tip becomes tapered. At this stage, the fiber also shows spiral growth and no longer grows toward micropyle. The fiber surface is coated with a lamellar cuticle which stretches and thins as fiber elongates.

All epidermal cells except the stomatal guard cells have potential to develop as fibers but fiber density is about 3300 per mm^2 and one out of four develop as fiber. Both lint and fuzz fibers originate as epidermal outgrowths of fertilized ovule. The fibers that initiate elongation on the day of anthesis are destined to become lint whereas epidermal cells initiating elongation in subsequent about fourth to fifth days after anthesis, produce only fuzz fibers.

Fibers does not indicate a uniform proliferation over the whole seed. Certain morphological changes associated with the fiber differentiation occur at the chalazal end of the ovule sixteen hours before anthesis and additional cells undergo differentiation closer to micropylar end by ten to twelve hours pre-anthesis, differentiation of fiber initials is observed at chalazal end, and occasionally fiber initiation from the crest of funicule is reported.Structural alterations associated with early stages of fiber cell elongation occur quite rapidly following anthesis. These appear to be correlated with the formation of central vacuole, plasma membrane, primary cell wall and increased protein synthesis necessary for extensive cell elongation. A dilated portion of endoplasmic reticulum in close association with the tonoplast showing a highly fenestrated membranous network suggests its derivation from endoplasmic reticulum.

Formation of the large central vacuole begins at the base of the fiber in a very precise manner and occupies most of the cell volume by two days after anthesis. Thus, a thin rim of cytoplasm separates the vacuole and the cell wall during elongation, and the various organelles including the nucleus are concentrated in the fiber tip. Dictyosome involvement in both plasma membrane and primary cell wall formation is suggested from observations of similarities between dictyosome-associated vesicles, containing fibrils which appeared similar in morphology to fibrils found in primary cell wall and plasma membrane associated vesicles. The differentiating fiber cells are enlarged and possess an enlarged nucleus which is transposed from the original central position it occupies in the ovular epidermal cell to an electron-dense cytoplasm. There is release of phenolic substances from the vacuole and an increased number of ribosomes are present in elongating fibers at anthesis. Phenolic compounds like O-diphenols presumably inhibit IAA oxidase to allow an intracellular auxin level high enough to initiate fiber differentiation. In non-differentiating fiber cells, the phenols are retained in the vacuole. More numerous ribosomes and rough endoplasmic reticulum are observed in fiber cells than in adjacent non-differentiating epidermis. The single nucleolus found in cotton fibers enlarges following anthesis , shows segregation of granular and fibrillar components by one day after anthesis, develops a large "vacuole" thus appearing ring shaped, and occupies much of the nuclear volume by two days after anthesis. The fibrillar component is the first to receive newly synthesized RNA which later passes to the granular component. Nucleolar vacuoles are spherical inclusions of low density, which characterize active nucleoli and have roles in RNA transport. Nucleolar vacuolation during fiber growth indicates simultaneous output and neosynthesis of nucleolar material. Prominent nucleoli are not observed in nuclei after ten days of anthesis. Thus, the amount of ribosomes synthesized in very early stages of fiber elongation may subsequently determine the rate of elongation and thickness of the fiber during maturation

Cell Wall

Primary cell walls of cotton fibers are laid down by undifferentiating cells that are still growing and it is these primary cells that control fiber growth. Secondary walls are derived from primary walls in cells which are stopping growth and are differentiating into cells with specialized functions. The cell wall of fiber cell is composed of cellulosic microfibrils embedded in an amorphous matrix of non-cellulosic substances consisting mainly of pectic polysaccharides, hemicelluloses, and proteins. During growth of the cell, polymers of the wall interact and change resulting in alteration of their properties For about one-half of the extension phase, the fiber cell contents are bounded by a thin primary wall which undergoes irreversible plastic extension and grows as the fiber enlarges. Primary wall constitutes a thin sheath on the outer surface of the mature fiber. The secondary wall is deposited on the inner surface of the existing primary wall, The secondary wall is less hydrated and more compact than the primary wall. In a mature fiber, the secondary wall is about 94% cellulose and relatively thick providing the extraordinary strength required of a textile fiber. Primary wall and secondary walls may be distinguished morphologically. In the former, the cellulosic microfibrils are oriented at a fairly large angle with respect to the long axis of the cell forming a loosely interwoven network although some regions exhibit parallel arrangement. The secondary wall is formed by microfibrils that are more closely packed, lie more parallel to one another, and are oriented with a smaller angle to the long axis of the cell.

In cotton fibers, the endoplasmic reticulum, Golgi complex, and plasmalemma form a functionally integrated membrane system for the synthesis and transport of cell wall components. Polypeptides of cell wall proteins are synthesized on the rough endoplasmic reticulum, as are other proteins, however, wall polypeptides are then segregated by release into the lumen of the endoplasmic reticulum. The polymerization of the hemicellulosic and pectic polysaccharides from sugar nucleotides and glycosylation of the structural protein of the wall take place in the dictyosomes after which the products reach the plasmalemma via vesicles arising from dictyosome cisternae. The vesicles fuse with the plasmalemma and then open toward the exterior, releasing their contents to the cell wall by reverse pinocytosis. Endoplasmic reticulum plays a major role in the synthesis of fiber secondary wall polysaccharides and swelling of the endoplasmic reticulum cisternae occur. The coated vesicles which are often concentrated in the vicinity of the dictyosomes and in cortical cytoplasm just beneath the plasmalemma in plant cells, occur in developing fibers during the primary and secondary wall formation.

Freeze-fracture electron microscopy has been utilized to identify structures associated with cellulose microfibrils which might play a role in microfibril synthesis and assembly. The fracture face of the plasma membrane shows both randomly scattered particles and large granules of 250-350 A° in diameter which are correlated with the impressions of microfibrils in the membrane.The ultrastructural evidence indicates that microtubules near the inner surface of the plasma membrane are arranged parallel to the orientation of wall microfibrils. The granular cellulose-synthesizing complexes which have lateral mobility in the plane of the plasma membrane are "spinning" cellulose microfibrils and being guided on "tracks" composed of microtubules.The microfilaments made up of actin (a contractile protein) and cortical microtubules are involved in directed microfibril deposition. Long cellulose molecules that make up the walls are juxtaposed into bundles to form fibrils which spiral down the length of the fibers. The microfibrils are held to one another by extensive hydrogen bonding and vary from 10 to 30 nm in breadth depending upon the species. Each microfibril cross-section contains roughly 10^3 molecular chains of b-D-glucan which exist as extended chains with a two fold screw axis, arranged in an ordered manner within the microfibril. The complex microfibrillar framework in the walls gives the fibers a tensile strength approximately that of a steel wire of the same diameter.

The deposition of the fibrils within the walls, i.e., whether they are parallel to the long axis, transverse or random, has an important bearing on the physical properties of the fibers. Therefore, how the plant synthesizes b-1, 4-glucan chains and deposits them in the cell wall is of great signficance as the arrangement of these fibrils is very unusual; rather than exhibiting a continuous spiraling in one direction, they intermittently reverse their gyre. At these reversal points, the wall of the fiber is thin and it is here that the mature fiber twists as it dries and collapses into a ribbon when it is exposed to the air. The occurrence of twists in the fibers enables them to be spun into yarn. The mature fiber, thus, consists of a collapsed tubular cell with a very small lumen and thick walls composed of superimposed bands of cellulose, but having thinner places where the direction of the cellulose spirals was reversed.

Bio-chemical Changes

Mature cotton fiber contains on an average cellulose, 94.0; wax-like substances, 0.6; pectins 0.9; organic acids, 0.8; nitrogenous substances, 1.3; ash, 1.2; non-cellulosic polysaccharides, 0.3; and, unidentified substances, 0.9 percent of absolute dry substace. The younger fiber possess, lower cellulose content but greater amount of other admixtures and the moisture content.

Carbohydrates

Cotton bolls accumulate good amounts of carbohydrates, mainly glucose, fructose, and sucrose during development. During primary wall formation, sucrose which is the transport sugar, is inverted so and glucose and fructose are accumulated for later use in the synthesis of secondary cell wall. On the other hand, the sucrose content increased regularly until fiber maturity. There exist separate "storage" and "metabolic" pools of glucose in the fiber showing mutual exchange. Since the fibers are highly vacuolated, the storage pool is the vacuole.

The fiber cell growth depends on metabolic energy and the breakdown of sucrose rather than of hexose as the starting point of carbohydrate metabolism. The non-photosynthetic and non-gluconeogenic cells receive the bulk of their organic carbon as source of energy. Sucrose utilization takes place by the action of invertase or sucrose synthase, and by hydrolytic clearage yields D-glucose and D-fructose. The D-glucose containing nucleotides are subsequently utilized in a variety of biosynthetic reactions & D-fructose is phosphorylated to corresponding hexose-6-phosphate by hexose kinases. Soluble and wall-bound invertase activities tend to be higher during the period of rapid fibre elongation. The hexose substrates are required by elongating fibres for energy metabolism and biosynthetic reactions which are made available by the hydrolysis of imported sucrose. The cytoplasmic and wall bound acid invertase show positive correlation with fibre length. The interaction of cytoplasmic a-galactosidase with acid invertase contribute to maintain sugar level as sugar oligosaccharides are cleaed by a-galactosidase to produce reducing sugars. A close correlation exists between rates of cell growth in a variety of tissues and acid invertase activity. A direct conversion of sucrose to sugar nucleotide for cell wall synthesis could account for significant fraction of sucrose breakdown in maturing cotton fibers.

Cotton fibers contain b-glucanates such as exo (1-3)-b-D glucanase, exo (1-3)-b-D glucanase and (1-4)-b-D-glucanase, though their activities varied at different growth stages. The activities of cellobiase and b (1-3) glucanase are higher during rapid elongation and declined at later phase of growth. They hydrolyze cellobiose and callose (b 1-3, glucan) to provide glucose units. Higher activities during elongation explains the channeling of glucose for osmoregulation and energy production. b (1-3) glucanase play role in elongation of fiber through hydrolytic rather than transglucosylase activity, since the glucose provided to hydrolysis could re-enter the low molecular weight sugar pool and could be used for synthesis of cellulose. The callose pool of fibre undergo second turnover at the beginning of secondary wall formation to about eight times in order to deliver glucose molecules required for the synthesis of cellulose during the same time. The role of b-glucanase is unclear though endonucleases are involved in cell wall loosening at the primary cell wall elongation phase. The enzyme also cooperates with the exoglucanase in hydrolytic degration of cell wall at secondary wall stage.

The secondary wall cellulose deposition begins at time closely related to the final length which fibers will attain just in advance of cessation of elongation. This could be expected that such an increase in cellulose deposition would cause an abrupt cessation of elongation but this does not happen until the second rise in rate of cellulose deposition occurs. It is possible that the microfibrils continue to be deposited in a random or transverse orientation during this phase of deposition, thus, do not limit cell extension. Other possibility could be that cells at this stage are elongating solely by tip growth and are not limited by deposition which occurs behind it.

The sugar content of cotton fibers is associated with cessation of fiber elongation and initiation of secondary wall formation. The amounts of galactose, mannose, rhamnose, arabinose, uronic acids and non-cellulosic b-1, 3 glucan reached maxima at the end of primary wall formation or at beginning of secondary wall formation and decreased thereafter. Contrarily, the amount of xylose and cellulosic glucan increased upto fiber maturation. The contents of sucrose, fructose and glucose varied at different stages of fiber development. The first maximum for fructose and glucose corresponds to beginning of secondary wall formation. Glucose in the fiber exists either in storage or metabolic pool which show mutual exchange.

Proteins

High level of proteins is recorded in the fibre cell wall upto the end of elongation period, after that it decreased and then reached a second maximum at the end of secondary wall formation. Ontogenetic changes in esterase and alkaline phosphatase have been investigated in developing fibers. Stimulation of membrane associated polysaccharide synthetases by a membrane potential in developing fibres is observed. Oxidases such as IAA-oxidase, peroxidase, O-diphenol oxidase show lower activity during elongation phase and high activity during secondary thickening. A shift in redox balance towards oxidation could play a significant role in fibre maturity. Cotton fibers possess several gluconases which appear associated with cell wall. The a-gluconases complex associated with cotton fiber cell wall also function as a transglucosylate introducing system.

Minerals and organic Acids

Cotton bolls accumulate minerals throughout their development. Redistribution of minerals (nitrogen, phosphorus and potassium) among bur, seed, and fiber may account for many of the compositional changes in each component especially during the period of boll maturation. This suggests the existence of a physiological continuum among the boll components during development. Ascorbic acid content is high in young fibers and decreases at maturity. Analysis of organic acids from the fiber shows malate and citrate to be the predominant ones. There is a measurable amount of water-soluble arabinogalactan-like polymers in the fibers.

Lipids

Mature fibers contain 0.5% of a lipid which is a mixture of waxes, fats, and resins. The lipid content in young fibers, is quite high. The fibers undergoing active extension incorporate most of the label from [1-^{14}C] acetate into polar lipids as compared with the non-polar lipids implying active membrane bio-sysnthesis. Lipid synthesis contributes to the tonoplast enlargement and thereby generation of turgor pressure. Apparently, the amount of lipid synthesized during fiber growth functions mainly in the synthesis of membranes and maintenance of their biochemical integrity. The presence of lipid bodies, sterols, steryl glucosides, esterified steryl glucosides, glucosyl-phosphorylphenol, fatty hormones, etc. has been ascertained in developing fibers.

The mean lipid content in mature fibers is 0.6 per cent. Sterylglucosides and acylated steryl glucosides are formed through transport of ^{14}C-glucose from UDP-^{14}C-glucose to endogenous sterol acceptors. The fibers undergoing active extension incorporate most of the label from 1-^{14}C acetate into polar lipids as compared with the non-polar lipids implying active membrane biogenesis. The amount of lipid synthesized during fiber growth functions mainly in the synthesis of membranes and maintenance of their biochemical integrity. b-sitosterol is predominant while co-campersterol, cholesterol and stigmasterol are present in smaller quantities. Palmitate and oleate are the major acyl components of esterified glucosides. The steryl glucosides and their acylated derivatives are found in membrane and their role in membrane permeability is suggested.

A particulate enzyme system from cotton fibers forms both steryl glucosides and acylated steryl glucosides by catalyzing the transfer of [^{14}C] glucose from UDP-[^{14}C] glucose to endogenous sterol acceptors. Analysis of the products by gas-liquid chromatography and mass spectrophotometry revealed that a-sitosterol is the predominant sterol moiety. The radioactivity first appeared in the steryl glucoside and then in the acylated steryl glucoside which is the immediate precursor of the acylated steryl glucoside. Phosphatidylethanolamine has been shown to be the best acyl donor by demonstrating that ^{14}C-labeled fatty acids from ^{14}C-labeled phospholipids can be transferred to steryl [^{3}H] glucoside to form a ^{14}C, ^{3}H-labeled acylated steryl glucoside. Colorimetric determinations of proline and hydroxyproline in developing fibers demonstrate their presence in the wall, and soluble fractions. The hydroxyproline content in fiber walls of a short staple cultivar is higher than its long counterpart during the period of rapid expansion. However, the absolute amounts of hydroxyproline detected in the walls may not be the limiting factor for cell extension. It is probable that the degree to which the fibers will attain their final length is contingent upon the time and mode of deposition and/or functional relationship of certain hydroxyproline containing protein(s) to other wall components in elongating cotton fibers.

Adenine nucleotides

The content of adenine nucleotides and energy charge requirements vary in *Gossypium* genotypes. In short staple cotton, the rate of dephosphorylation of ATP was lower relative to long staple. Long fibers utilize more ATP and had high energy requirements as compared with short fibers. Cellulose synthesis in fibers can be limited by low supply of high energy phosphates so keeping the cells adequately supplied with this energy is one way of regulating cellulose synthesis. Increase in level of glucose-6-phosphate, 6-phosphogluconate, fructose-6-phosphate, phosphoenolpyruvate and pyruvate occur concomitant with time of rapid fibre growth so as to cope with it.

Nutrient and Metabolite requirement for in vitro *culture*

Cotton ovule culture has been used to study the role of micronutrients in fiber growth. A constant supply of boron is necessary to maintain fiber elongation and prevent callusing of epidermal cells *in vitro*. In the boron-deficient medium, ovules produce callus and accumulate brown substances. Profuse callusing in the absence of boron occurs only in the presence of GA_3. Thiamine is the critical vitamin essential for GA_3-induced callus formation when unfertilized ovules are cultured in the absence of boron. Boron deficiency-like symptoms are induced by 6-azauracil (inhibitor of orotidine monophosphate decarboxylase) in ovules growing in boron-sufficient medium. The correlation is further strengthened by the finding that orotic acid and uracil partially overcome both boron deficiency and azauracil effects. Boron deficiency symptoms are probably related to reduced activity in the pyrimidine biosynthetic pathway which may cause reduced synthesis of UDP-glucose and other UDP-sugars involved in cell wall composition of the fiber. Total incorporation of [6-^{14}C] orotic acid into fiber was inhibited under boron deficiency, indicating that effect may be at the membrane transport level or at an early stage of orotic acid metabolism i.e. inhibition of orotodine monophosphate decarboxylase.The percentage incorporation into UDP- glucose is lower under boron deficiency and also its incorporation into cell wall material of fibers. UDP-glucose pyrophosphorylase result in the observed decreased radioactive orotic acid incorporation. In sum boron deficiency in cotton fibers causes a general inhibition of orotic acid incorporation whereby pyrimidine synthesis intermediates are shunted away from UDP-glucose synthesis and channeled preferentially into RNA synthesis. This could be related directly to cessation of fiber growth due to inhibition of wall synthesis. NH_4^+ is another important factor in the growth and development of cultured cotton ovules. The ovules cultured at 28°C require IAA and either NH_4^+ or GA_3 in the medium for fiber development whereas ovules cultured at 34°C require only IAA. NH_4^+, at increased temperature has no effect on the induction of increased nitrate reductase activity in the ovules so that the effects of these compounds on fiber development are independent of the availability of reduced nitrogen as a general substrate for growth.

Respiratory Changes

Fiber cell growth depends upon metabolic energy available. During carbohydrate oxidation, the energy stored in carbohydrate molecules is utilised for the endergonic activities of cells. At the same time, the metabolism of carbohydrates provides a number of intermediates for biosynthetic processes and cellular maintenance. It is probable that non-photosynthetic and non-gluconeogenic plant cells such as fibers receive the bulk of their organic carbon as sucrose and hence the breakdown of sucrose rather than the metabolism of hexose is the starting point of carbohydrate metabolism in these cells. Sucrose utilization is initiated by two enzymatic reactions : hydrolytic cleavage to D-fructose by the action of invertase, and/ or cleavage by sucrose synthetase to produce fructose and sugar nucleotide intermediates, primarily UDP-glucose. High acid invertase ensures a large net demand of hexoses for extensive fiber extension. The presence of cell surface located invertase for cotton fibers and one of its possible roles in regulation of cell wall synthesis by mediation of uptake of sucrose from the apoplast into the symplast has been suggested. The reaction of hexokinase funnels hexoses into intermediary metabolism by phosphorylating glucose and fructose to the corresponding hexose 6-phosphate which may undergo glycolysis, enter the pentose phosphate pathway, participate in oligo- or polysaccharide synthesis, or be hydrolyzed to free hexoses by phosphatases. The hexokinases, particularly, those which display high activities toward D-fructose, are most important in the metabolism of rapidly developing non-photosynthetic cells which have large requirements for precursors of cell wall polysaccharides. UDP-glucose is enroute to the synthesis of cellulose in developing cotton fibers. Thus, a direct conversion of sucrose to sugar nucleotides for cell wall synthesis could account for a significant fraction of sucrose breakdown in cotton fibers. It is likely that the rate of sucrose import and growth of fibers may be controlled by an interplay of invertase and sucrose synthetase which ensure that the rate of import is precisely matched with the rate of its metabolic utilization.

Glycolysis and the pentose phosphate pathway operate in elongating cotton fibers and the extent of their operation varies with the demand for respiratory products. Hexokinase, glucose-6-phosphate dehydrogenase, phosphofructokinase ; pyruvate kinase, and succinate dehydrogenase show increased activities during the period of rapid extension growth and decreased activities when the rate of growth slows down. The oxidation of [1-^{14}C]- and [6-^{14}C] glucose and measurements of important glycolytic and pentose phosphate pathway intermediates yielded a similar pattern. The increased channeling of metabolized glucose into the two pathways during the period of active fiber growth reflects a high requirement for energy and reducing power which must be produced to attain cell extension of considerable magnitude. Cotton fibers have high levels of adenosine phosphates and low energy charge during this period. As the rate of fiber growth slows down, the decline in enzyme activities, metabolites, and turnover rates of [^{14}C] glucose points to a shift in metabolic priorities. Over 60% of the translocated sugar accumulates in wall polymers, a high proportion of the carbon passing through the metabolite pool of glucose is used for the synthesis of cellulose and b, 1-3-glucan alone. However, a significant level of glucose oxidation continues to support the on-going fiber development.

The pentose phosphate pathway is functionally more important in non-photosynthetic cells to compensate for the NADPH, normally produced in the chloroplasts. The cotton fibers are equipped with an additional pathway of NADPH generation. It is seen that when the activity of the pentose phosphate pathway decreases during the later stages of fiber growth, malate accumulated via CO_2 dark fixation is catabolized to supply NADPH via decarboxylation by malic enzyme. An increase in the activity of the pentose phosphate pathway may reduce the rate of malate decarboxylation or vice-versa through the regulation of the $NADP^+$/NADPH ratio.

All the three forms of high energy molecule (ATP) are detected in cotton fibers. The rate of fiber elongation may be the result of ATP levels. One of the most important cellular "sinks" for ATP is attributable to the regulation of ionic fluxes. The role of ATP in maintenance of cell integrity, regulation of cell turgor, phosphorylation of metabolic substrates, and in other energy utilizing biosyntheses, such as the synthesis of RNA, proteins, membrane lipids, and nucleotide sugars could potentially influence the rate of fiber growth in many ways. The elongating fibers had relatively low energy charge during the period of rapid cell growth. The cell growth is an energy-consuming system rather than an energy-yielding or storage one. Therefore, energy charge may be controlled by the rate of ATP usage whereby low values of energy charge are expected during rapid cell growth.

The activity of phosphofructokinase and pyruvate kinase *in vivo* may be regulated by the rate at which ATP is used. It is pertinent to note that the ATP/ADP ratio is markedly lower in the long fibers relative to the short counterparts which may be responsible for the respiratory augumentation of the former. A faster hexose consumption by long fibers has physiological relevance for attaining increased length whereas a metabolic depression in terms of hexose oxidation in the short fibers may dwarf their growth.

Dark fixation of Carbon Dioxide

Cotton fiber grows in the dark interior of the boll protected against the environmental perturbations. Presumably, cotton boll growth results in the production of elevated levels of CO_2 inside the boll. One of the major effects of increased CO_2 concentrations is to increase its rates of fixation catalyzed by phosphoenolpyruvate carboxylase. Developing cotton fibers possess an active system for assimilating CO_2. Studies on unfertilized cotton ovules show that fiber growth is inhibited by the absence of K^+ and CO_2 in the culture environment. The fiber growth is dependent on the turgor pressure in the fiber and that K^+ and malate are the osmotically active solutes (osmolytes) which are largely responsible for the production of turgor in the fiber. Malate is partly synthesized by dark CO_2 fixation.Maintenance of turgor or pressure potential is mandatory for continuous cell expansion which is achieved by increasing the number of osmolyte molecules in the cell.. In elongating cotton fibers, K^+ and malate levels fluctuate in correlation with the growth rate and reach peak levels when the growth rate is maximum. Glucose and sucrose are the predominant sugars in fibers and "vacuolar" and "cytoplasmic" pools have been observed. Studies on vacuoles from sugarcane suspension cultures show that tonoplast energization may play a decisive role in both active hexose uptake and active sucrose uptake at the tonoplast. In a need for carbon economy, the cotton fibers seem to have successfully exploited the dark metabolism of CO_2 as an adaptation for intracellular recycling of CO_2 by refixation to achieve osmoregulation of growth and to fulfil specific metabolic requirements. Malate produced by dark fixation pathway possibly acts as an osmoticum and a counterion for K^+ accumulation during the period of active fiber extension. The enzymes of malate metabolism, i.e., phosphoenolpyruvate carboxylase (PEPc), glutamate-oxaloacetate-transaminase (GOT), NAD^+ -malate dehydrogenase (MDH), and $NADP^+$ -malic enzyme (ME) are readily detected in both *in vitro* and *in vivo* growing fibers. During the period of rapid growth *in vivo*, the fibers contain enhanced activities of PEPc, GOT, and MDH which decline afterward to low values when the rate of growth slows down. PEPc directs a portion of the glycolytic carbon in the form of PEP toward the formation of C_4 acid, oxaloacetate, which once formed can either be incorporated into citrate, transaminated to aspartate, or reduced to malate. In rapidly elongating fibers, reduction of oxaloacetate to malate is the main route of the metabolism of fixed carbon. This is evident from the elevated levels of MDH compared with that of GOT, low and unchanged activity of ME. Malate accumulation in elongating fibers reaches a maximum when the growth rate of fibers is highest..

K^+ is essential as an activator of many enzymes in key metabolic processes like glycolysis, tricarboxylic acid cycle, oxidative phosphorylation, RNA and protein synthesis.The regulation of turgor pressure is related to the vacuolar system. As the turgor increases with increasing osmotic pressure in the vacuole, the malate housed in vacuole leaks out into the cytoplasm Control of tonoplast influx of other ions by vacuolar concentration or cell turgor is well established. When the fiber growth slows down a dramatic increase in ME activity coupled with a distinct decrease in fiber PEPc activity and malate content is observed. Malate efflux from the vacuoles is caused by changes in membrane permeability of the tonoplast, a decrease in cellular pH, mineral ions like Mg^{2+}, Ca^{2+}, and changing $NADP^{+}$/ NADPH ratios.

At later stage of development , a large proportion of the carbon supply to the fiber is used for the synthesis of a thick secondary wall and glucose oxidation decreases. MDH and ME together constitute a transhydrogenase system converting NADH to NADPH. Pyruvate is produced concomitantly with NADPH by ME reaction. In this respect, the dual role of malate as an osmoticum and a carbon-energy source during fiber growth seems feasible. Malate is a useful form of stored carbon in plant cells and can be tolerated at quite high concentrations in the vacuoles.

K^{+}, and malate were found to be higher in elongating fibers of a long staple than short staple cotton. This is supported by a faster turnover of [^{14}C] glucose in the long fibers and thus possess efficient CO_2 production and fixation systems relative to short fibers The ability to generate malate in this way seems to be one important implication for osmoregulation of growth. It was also observed that during the period of active elongation, the K^+/malate ratio is higher in the *arboretum* cotton than the *hirsutum* cotton. High K^+/malate ratios of short fibers are mainly due to the lowered malate levels of these fibers. Phosphoenolpyruvate is a major branch-point in glycolysis and carboxylation of this intermediate by PEPc is important in CO_2 and NH_4^+ assimilation to provide a source of respiratory substrates, amino acids, counterions and NADPH for cell growth. Each complete turn of the tricarboxylic acid cycle results in the indirect oxidation of one acetate molecule and the release of two CO_2 molecules. Thus, the number of carbon atoms lost as CO_2 matches the number of carbon atoms entering the cycle as acetate, and if this were the only loss the cycle would be self-sufficient. However, various intermediates are drained from the cycle and used in biosynthetic reactions of cell growth. Carbon removed in this form must be replenished for the cycle to continue its operation. Anaplerotic carboxylation of phosphoenolpyruvate probably operates and helps to meet this demand.

Microtubules lining the periphery of the protoplasm tend to parallel the deplovment of cell wall microfibrils and the pattern persists. Cell wall synthesis is a developmentally regulated event. The cell wall of the cotton fiber is a dynamic structure as its composition is continuously changing throughout development ending with the cessation of the fiber's metabolic activity. The primary cell wall is an amalgam of a dynamic equilibrium of component polysaccharides during development but always has a relatively low cellulose content. The actual amount of cellulose in primary walls of elongating fibers is a constant value. Cellulose deposition sharply rises at the onset of secondary wall deposition, it declines, and then rises again. The primary wall is made up largely of pectic substances and a decrease in the percentage of pectin is recorded with an increase in the age of the fiber, the rate of decrease being more rapid when the secondary wall is laid down. The kinetics of the cell wall was relatively constant until about day twelve when it increased markedly until secondary wall cellulose deposition was completed. Between days 12 to 16 after anthesis, increases in essentially all the components contribute to increase in total wall thickness. At the time of onset of secondary wall cellulose deposition, a sharp decline in protein and uronosyl residues occurs. After day 16, deposition of cellulose are the only events which contribute to continuing increase in wall thickening. The deposition of wall thickening takes place in layers that are alternately denser and lighter under normal day-night conditions, but are absent in fibers developing under constant light. The alternating structure gives tensile strength to fibers.

The absolute amounts of frucose, galactose, mannose, rhamnose, arabinose, uronic acids, and non-cellulosic b-1, 3-glucan all reached a maxima at the end of primary wall formation or beginning of secondary wall and thereafter decreased, implying that degradation of non-cellulosic polysaccharides was occurring. In contrast, the amounts of xylose and cellulosic glucan increased until the end of fiber development. A steady decrease in the percentage of protein content of the walls is recorded with increasing age. Both soluble and insoluble forms of b-1, 3-glucan increase in relative amounts upto the time of onset of secondary wall formation. Viscometric methods to study derivatized of glucan chains suggested that biosynthesis of cellulose consists of two distinct kinetic stages corresponding to the formation of primary and secondary walls during the development of the cotton fiber. The first stage proceeds very slowly and yields a small amount of "primary" cellulose having a non-uniform degree of polymerization (DPw) ranging from 2000 to 6000. The second stage proceeds much faster and provides a large amount of "secondary" cellulose having a DPw of 14,000. During the second stage, the DPw is independent of variations in the kinetics of the rate of synthesis of cellulose.

The orientation of polar glucose molecule when it enters plasma film for polymerization into cellulose is influenced by the intensity of polar forces of the lint cell. The intensity of electrical field varies both with the level of supply and rate of utilization of carbohydrates. Sugar nucleotides are the most likely activated monomer substrates for cell wall polysaccharide synthesis. In cotton fibers, the UDP-sugars are the predominant nucleotide sugars whereas GDP-sugars are not detected in significant quantities. The extent of incorporation of UDP-glucose increases with increasing age of the fibers and was found to be very high in fibers from 20 to 30 days after anthesis. Also GDP-glucose pyrophosphorylase activity is undetectable but UDP-glucose pyrophosphorylase activity is found in extremely high levels throughout fiber development The predominant products synthesized from UDP-glucose by extracts of cotton fibers are sucrose, steryl glucosides, b-1, 3-glucan, and cellulose and the rate of synthesis and turnover of UDP-glucose is more than sufficient to account for the combined rates of accumulation of these constituents. The synthesis of glucolipids from UDP-glucose with the biochemical properties of glycosyl phosphoryl-polypenols was reported.Cotton fibers can also incorporate mannosyl portion of GDP-[^{14}C] mannose.The lipid-linked saccharides participate in the synthesis of glycoproteins, mannan or glucomannan, and cellulose or other glucan synthesis. Glucolipids and again oligosaccharide-linked lipids were formed from the substrate UDP-glucose and endogenous lipids.

Cell walls from growing cotton fibers contain appreciable amounts of non-cellulosic glucans. Hydrolysis of these glucans by different types of glucanases show that they are essentially composed of a-1, 3-linked residues. There are no indications of mixed linked b-1, 3/b-1, 4-glucans. The first layer of secondary wall has been referred to as the winding layer in fibers and most of b-1, 3-glucan is associated with this layer. As fiber elongation continues during the period of deposition of the winding layer, there seem to be differences among cotton varieties in the extent of elongation which occurs after the onset of secondary wall formation. UDP-glucose b-1, 3-glucan synthetase has been reported in the fibers which are often activated *in vitro* by the substrate UDP-glucose and by b-linked disaccharides. Callose is an integral but possibly transient component of fiber cell walls. At secondary wall formation, callose is synthesized at a very high rate but its total amount in cell wall does not exceed 2 percent in later stages of development. Callose is an intermediate in cellulose synthesis at the stage of secondary wall formation and may accumulate where biosynthetic step is inhibited. Fluorescence microscopy, after staining with aniline blue revealed that callose is always localized independently of the fiber age in the innermost wall layer bordering the cell lumen; from the onset of the secondary thickening until the end of fiber development. Incorporation of radioactivity from [U-^{14}C] glucose into the cell wall polymers 2,6-Dichlorobenzonitrile (DCB) was most effective and specific one for its effect on cellulose synthesis. Coumarin, generally regarded as an inhibitor of cellulose synthesis in other plant systems, was effective in millimolar concentration and, like DCB, was relatively specific with regard to its effect on cellulose synthesis. Transmembrane potentials were induced in membrane vesicles isolated from fiber by establishing artificial diffusion potentials with combinations of anions and cations having different permeability coefficients. A potential, positive with respect to the inside of the vesicles, was established by suspending vesicles in 50 m*M* K^+ in the presence of the K^+ specific ionophore, valinomycin and the relatively impermeant counterion 2-(*N*-morpholino) ethanesulfonic acid. Valinomycin can transport K ions across the vesicle membrane, down the concentration gradient thereby establishing a net positive (inside) potential. The induction of a positive (inside) potential in a membrane preparation from developing fibers resulted in a substantial (4- to 12-fold) stimulation of the activity of membrane associated a-glucan synthetases which produce b-1-3&1-4glucan.but not a-1, 4-glucan (starch). The stimulation of activity is the consequence of a transmembrane electric potential and is not influenced by a pH gradient. The variability in fiber length and secondary wall thickness of fibers among different cotton varieties depends upon the interplay of a complex pattern of metabolic processes and regulating mechanisms that govern fiber growth.

□□□

13

Improvement in Cotton Fiber through Biotechniques

Advances have been made in understanding the development of fiber, wall characteristics and biochemical changes accompanying growth of fibers. A new set of biotechniques are unfolding complexities of fiber cell walls which are responsible for significant physical traits of cotton fiber. Genes from other sources have been introduced into cotton to improve insect and herbicide resistance. Also genes coding for cellulose biosynthetic enzymes from bacteria have been introduced. Wrinkle resistance and cotton having desirable dye-uptake potential and variable natural shades are new blockbusters of this technique. Thus, available biotechniques are able to permit an unprecedented ability to modify cotton fiber quality.

Developing cotton fibers provide an excellent model system to investigate processes of cell elongation and wall development. It has been possible to develop protocols to culture cotton ovules/seeds at desired stages in defined basal medium. From the surface of the ovule relatively synchronous population of highly uniform cells is obtained and the system has been exploited to unravel several problems e.g. the developmental sequence of the fiber; factors regulating ultimate length of the fiber and the transition between primary and secondary wall synthesis. Cotton fiber differentiation and development involve several biochemical, physiological and morphological alterations. The phenotypic expression is determined by genotype, though the development of fiber is regulated by a combination of internal stimuli, hormonal contents and external influences. Most studies dealing with fibers have been concerned with the two main phases of development, elongation and secondary wall deposition.

In general two approaches so far have been adopted in cotton regarding hormonal effects and these are : a) isolation, identification and quantification of hormones from bolls, ovules and fibers. Several workers have isolated gibberellins, auxins, cytokinins, abscisic acid and ethylene from cotton fruits and/or ovules having attached fibers at different stages of boll development. The second approach, involves external application of hormones *in vitro* as well as *in plantae*. Exogenously divergent plant hormones have been sprayed on the cotton plants at different stages of development to influence boll growth and fiber quality. For this several techniques have been used and these are spraying the whole plant, soaking the ovaries, or injecting the desired hormones in the bolls. Under field conditions cotton plants respond differently to growth regulators than to trials in greenhouses or growth chambers. In any case extending the period of fiber elongation or secondary wall thickening and stimulating the translocation of photosynthates to the fruits through growth regulators was sure to enhance cotton production and improve the quality of fiber.

Gibberellins and auxins are important factors for fiber differentiation. Hormones or their inhibitors are added in the culture medium to establish their role. SEM studies of the cultured ovules have been undertaken and role of polyamines and phenolics has been suggested. Ortho-diphenol oxidase activity during fiber initiation and its role in shifting redox balance towards oxidation possibly plays a significant role in fiber initiation. The essentiality of IAA for fiber initiation and GA for the expansion of ovules has been demonstrated using metabolic inhibitors of two hormones through the estimation of IAA oxidase or peroxidase activities as well as through SEM studies. *In Plantae,* pollination seems to trigger hormone induction whereas *in vitro* unfertilized ovules required the addition of the two hormones. The significance of exogenously added hormones seems to be dependent on whether the ovules used in culturing were fertilized or not. In the fertilized ovules, only GA is needed to support ovule growth. Kinetin did not support fiber growth and induced callusing in the ovules. Thus, a precise balance of hormones was required to ensure proper fiber development. During fiber initiation and elongation the nucleus and nucleolus undergo marked changes in size and structure pointing towards enhanced capacity for ribosome synthesis. Studies on effect of GA, IAA and ABA alone or in combination have shown to affect the nucleolar evolution and concomitant fiber length especially in cultured unfertilized ovules. The input of all the three hormones in a balanced fashion seems to be important for the regulation of synthesis and release of nucleolar material.

Fiber development in cool temperature

Due to tropical habitat cotton is highly sensitive crop to cool temperature. Temperatures less than 16°C essentially inhibit fiber growth. Cool nights result in suboptimal cotton yield and reduced fiber maturity. Night temperature (10, 15, 20, 27°C) regimes affected elongation of fibers in field grown *hirsutum* cv. Fiber age and night temperature determined rate of fiber elongation, and temperature coefficient of elongation decreased with progressive fiber age and night temperature. The initial stages of fiber elongation are highly temperature sensitive, while the later phases (elongation and secondary wall formation) were temperature independent. Early delay in fiber initiation by cool temperature extended the elongation period. On the basis of the response of fibers to cool temperature, three stages of fiber development could be recognized; fiber initiation, very early elongation and subsequent elongation. Of these, the first two stages were highly sensitive to cool temperature. Fibers of plants exposed to adverse diurnal temperature fluctuation elongated at reduced rates for an extended period but attained the comparable final length to those grown under favourable conditions.

The adverse response of cotton fiber development to cool temperatures was ovule/fiber specific event. Low temperature altered fiber development even though sufficient carbon source was available. Incorporation of (U-^{14}C)-sucrose into callose and cellulose after one hour was 79 and 21 percent, respectively.

Cellulose synthesis and respiration responded differently to cool temperature. A decrease in temperature from 28°C to 15°C reduced respiration and cellulose synthesis and increased roughly in a linear manner between 18°C and 28°C. The rate of cellulose synthesis plateued upto about 37°C and dropped off at 40°C, suggesting that more energy was released through respiration above 28°C but was not necessarily used for increased cellulose synthesis.

The pattern of fiber initiation was affected by cool temperature. Cotton ovules by two days pre-anthesis possibly acquired an internal signal that permitted initiation only on the normal day of anthesis. Initiation and early fiber elongation were found as two different physiological processes and were independently affected by cool temperatures. Similar developmental response of *in vitro* cultured ovules to cool temperature has been shown. Ovules subjected to 32°C/15°C cycling temperature had delayed initiation of ballooning by one to two days followed by slow initiation and fiber elongation. Once the fibers attained a length of nearly 0.4 mm their further elongation was independent of temperature effect. Furthermore, *in vitro* culturing of ovules with fibers 0.4 mm length at 32°C/15°C cycling and 32°C constant had comparable rates of fiber elongation. The delay in fiber development induced by cool temperature could be overcome by the supplementation of guianolide derivative-I into the culture medium. This compound caused preponderance of bulging of epidermal cells & was highly promotive.

Cool temperature also affected biochemical processes. Sucrose and glucose levels decreased in ovules cultured in cool temperature even though glucose was not a limiting factor. The cycling effect of temperature (32°C/22°C) on sugar accumulation was reported up to the phase of fiber elongation. During secondary wall thickening even at higher concentration glucose was not used efficiently to synthesize and deposit secondary wall around the fibers. Seemingly, low temperature does not affect availability of glucose during secondary wall thickening phase though primordial initiation was delayed by the limitation of glucose. The low temperature has independent effects on the initiation, elongation and secondary wall deposition process. Cool temperature adversely affected the activities of a-glucosidase, acid invertase and o-diphenol oxidase at three mentioned stages with marked decrease in the activities of these enzymes during initiation and early elongation phases. Cool temperature markedly decreased the activity of o-diphenol oxidase and this in turn caused irregular and delayed initiation of fibers. The role of o-diphenol oxidase pertinent to auxin turnover and redox balance of developing cotton fibers was suggested. Cool temperature reduced the activities of oxidases in one day pre-anthesis ovules and their subsequent increase till five days post anthesis is correlated with essentiality of auxin during fiber initiation. The activity pattern of o-diphenol oxidase indicated a shift in redox balance from a reduced to an oxidative state in developing ovules. O-dihydroxyphenols are implicated in redox potential pertinent to cell division (reduction), cell differentiation and senescence (oxidation). The low o-diphenol oxidase activity with guaianolide derivative compared with the level during the initiation phase suggested that dihydroxy phenols maintained-SH group in a reduced state hence aided initiation and early elongation of fibers. A growth retardant (E)-1 (4-chlorophenyl)-4, 4-dimethyl-2[1,2,4, triazol (-1y-1)] pentene-3-ol (S-3307) is an antigibberellin and completely suppressed fibers on ovules suggesting role of GA in differentiation of initials. Considerable decline in the activity of acid invertase by S-3307 is indicative of the role of GA in regulating the activity of acid invertase and hence high apoplastic hydrolysis by preventing sucrose accumulation.

In developing fibers both intercalary or tip growth processes are operative. Several factors regulate the ultimate length of the cotton fiber, the extent of secondary wall thickening, and the transition of primary wall and secondary wall synthesis ultimately determine the quality of the fiber.

In-vitro culture techniques

Cotton has been in the forefront in several areas of *in vitro* culture techniques including ovule and cell culture and regeneration of plants from cells. These developments had enabled cotton to be genetically engineered. Cell and organ culture have been used for basic studies on cotton physiology and genetics. Although great progress has been made in the field of improvement of cotton with conventional breeding but, it is time-consuming and commercialization of new cotton varieties, often takes 6 to 10 years. Compatibility limitations narrow the gene pool available for this process. A number of these shortcomings may be overcome by plant tissue culture. For example, control can be exerted over selection of the gene(s) and its expression. The gene pool can be expanded to all living organisms (plants, animals, bacteria and fungi). As technology is refined, custom made synthetic genes will become another source for desired traits, so these techniques can significantly be applied for the improvement of cotton.

Ovule and Seed Culture

Conditions for culture of pre- and post-anthesis ovules/seeds from cotton have been established. The cotton ovule culture system is used extensively in studies pertaining to cellulose synthesis during fiber developments, cytoskeletal elements during fiber development, activities of various enzymes, calcium mediated changes in the activities of enzymes, influence of hormones on ^{14}C-partitioning and to cool temperature effects. Fibers grown on *in vitro* cultured ovules/seeds have the same chemical composition as fibers grown *in plantae*. However, *in vitro* fibers attained lengths nearly onehalf of that reached by *in plantae* fibers. Cells released from cotton ovules in suspension cultures have also been raised *in vitro*. The single cells elongated continuously in culture medium and attained a length of 2-3 mm. Studies are directed towards mimicking the ovule/seed culture and hormonal trends noticed *in plantae*. Since, Beasley and coworkers developed an ovule culture system to study the development of cotton fibers, studies using their system have greatly improved the understanding of the physiology of fiber development. Scientists have obtained cotton fiber, which was as long as the fiber *in vivo*, though strength was not comparable to the one produced *in vivo*.

Another application of cotton ovule culture is to obtain hybridization through the culture of the young ovule to maturity of the enclosed embryo. The culture system was subsequently applied successfully to the wide hybridization of wild *Gossypium* species with cultivated cotton. This technique overcame the physiological barriers to most crosses, and as a result, the diversity and availability of germplasm for cotton improvement was greatly increased. Several institutes used ovule culture to introduce fine characteristics of wild species into upland cotton, such as: pest resistance, disease resistance, resistance to adverse conditions, and glandless seed/glanded plant trait. Some of these were successfully bred. Through the use of culture system it has been demonstrated that cotton ovule tissues, like exposed tissues, possess defence mechanisms against fungal attack. *In vitro* technique was used to recover germplasm material from its native habitat when seeds are not available during collection. Diploid Asiatic *Gossypium* species are being hybridized to introduce pest resistance; and produce bridge species (D genome), which are used to facilitate genetic recombination with upland cotton. Additionally, several genetic nuclear traits are being transferred from wild species to cultivated ones.

Shoot and meristem culture

Shoot and meristem culture is an easier method to obtain regenerative plants. Scientists have developed shoot and meristem culture for genetic engineering and conservation of cotton germplasm materials. Transgenic cotton plants have been obtained by transforming shoot meristems by the method of particle bombardment. Meristem culture of cotton is being used for transformation with *Agrobacterium*. It has advantage over cell culture for genetic engineering since plants can be obtained from any genotype rather than being limited to those genotypes with regenerative ability (Coker lines).

Cell culture

Cells may be cultured as cell suspension or callus. The suspension cultures are widely preferred to investigate basic biochemical mechanisms and to raise somaclonal variants. The mechanism of elicitation of phytoalexin production was studied in cultured cells. In cotton phytoalexin are sesquiterpene compounds and H_2O_2 is involved as a second messenger in the stimulation of phytoalexin production. Suspension cell culture is also used to study the interaction of *Xanthomonas malvacearum* with blight resistant lines of cotton. Development of single cells into fiber like structures has been achieved in suspensions derived from callus of cultivars including the embryogenic line Coker 310, through repeated culturing and exposure to *Phymatotrichum omnivorum* toxin.

Somaclonal variants have been used to select cell populations tolerant to a variety of stresses. Cells tolerant to low and high temperature have been selected and plants have been regenerated. The ability to regenerate into plantlets was genotype-specific and was heritable as the quantitative trait.

Somatic embryogenesis and plant regeneration

Somatic embryogenesis and plant regeneration are fundamental to the genetic improvement of cotton using biotechnology and genetic transformation. A well-defined, reproducible, and highly efficient somatic embryogenesis and plant regeneration scheme is a pre-requisite for genetic transformation because plant cell transformation is a relative rare event.

Somatic cell culture

The regenerative plant was obtained from cotton somatic cell culture via somatic embryogenesis. Some important aspects and factors, such as explant, genotype, medium, hormones and their effects on cotton somatic embryogenesis and plant regeneration have been studied. Based on this, an efficient protocol for cotton tissue culture and plant regeneration has been established and whole plants could be regenerated within 60-80 days.

Somatic cell culture has been used to select somaclonal variants. It is an excellent system to select the variations against adverse conditions, such as salt stress, low temperature stress, or toxins produced by *Fusarium* or *Verticillum* wilt. Callus derived was exposed to stepwise levels of toxins of *Fusarium* or *Verticillum* wilt and salt concentrations. After 20-30 days of culture, the surviving calli were transferred into the next higher concentration. After 3-5 times of selection, somaclonal variations resistant to *Verticillum* wilt or NaCl were obtained.

Various physiological or genetic variations existed in the regenerative plants and their progenies, but the physiological variations converted to normal after transplanting into the soil. In cotton somaclonal variants, the leaf and fertility variations were always physiological, while most of the variations of plant shape, maturation, yield characters and fiber quality characters were genetic. Most of the variations were disadvantageous to breeding, but some of them could be applied in cotton breeding and production. Through continuous self-cross and screening, some new lines with high lint, high fiber quality have been obtained. The success has been achieved in encapsulating cotton somatic embryos and obtaining cotton artificial seeds. The germination capacity of the cotton artificial seeds under sterile conditions has reached 40 per cent.

Protoplast culture

Protoplast culture is required for obtaining somatic hybrid cells. In earlier reports only callus was obtained by protoplast culture but regenerative plantlets from protoplast culture of island cotton (*Gossypium barbadense* L.) were produced later on. The regenerative plants were obtained from protoplast culture of G. *hirsutum* L. and protocols for cotton protoplast culture have been improved.

Anther culture

Haploid callus and later on haploid plants were produced by embryogenesis and plant regeneration from cotton anther culture. The haploid plantlets were obtained from wild species, G. *lotzschianum* Anderss and regenerative plants from the callus of G. *hirsutum* L. cv. Coker 201, Siokral 1-3 and Lumian 6 via cotton anther culture.

Genetic engineering of cotton

Most accomplishments in genetic engineering have been made by several biotech companies, due to sufficient value of modified crop, large market value for the seed; engineer resistance into crop, exploitation of biological organisms for biocontrols, etc. Well characterized gene(s) having economic potential, technique for gene transfer and regeneration of target plant genotype from a single cell have been undertaken. *Agrobacterium*-mediated transformation dominated in most companies and public institutions, however, only Coker lines were used among the commercial cultivars. Transformation of meristems or germ cell lines permitted gene to be inserted into any commercial genotype without undergoing extensive backcrossing from a regenerated Coker line. Particle acceleration technology is used by several companies including Calgene, Monsento, Phytogen and Agracetus. A technique of particle acceleration has been tried in cotton using embryogenic cell suspension of Coker 312 as the target tissue. Co-transformation with two or more plasmids successfully has been achieved, transformation with as many as 11 plasmids at once has been attempted though selectable resistance gene was only on one of the plasmids. This technique has strong implications for rapid transformation of complete biosynthetic pathways into cotton.

The availability of genes that are thought to be potentially useful, if successfully engineered into a desired genotype, has determined direction taken in cotton biotechnology. Two areas have witnessed great strides :

(i) Resistance to broad spectrum herbicides.

(ii) Insect resistance based on production of toxins by *Bacillus thuringenesis*.

Transgenic cotton for insect resistance

Cotton is an intensely managed crop needing lots of pesticide per acre. It is infested by arthropod pests causing a loss in fiber production. Thus, engineering a pest-resistance cotton is a prime goal in any cotton improvement programme. Insect pests are a major problem in cotton production. Since the end of the 1980s, cotton production has experienced a decline in both yield and coverage area. The decline in yield of 15 to 30% has mainly been caused by bollworm (*Helicoverpa armigera* Hubner) infestation. Outbreaks of cotton bollworm infestation caused direct economic losses. Furthermore, farmers were discouraged from growing cotton. As a result, the national growing area has decreased.

Cotton plants are attacked by a number of insect pests, the major ones are the cotton bollworm, cotton pink bollworm (*Pectinophora gosspiella* Saunders), and cotton aphid (*Aphis gossypii*) which cause extensive damage if left uncontrolled. *H. armigera*, in particular, is becoming a serious threat as it is resistant to all of the currently available chemical pesticides, such as synthetic pyrethroids and organophosphates. Millions are spent each year on chemical pesticides to control cotton insect pests, a significant proportion of it being toxic organosulphur and organophosphorus insecticides such as endosulfan and pyrethroid. The more environmentally friendly synthetic pyrethrin insecticides have been effective in the past, but due to development of resistance by the insects may soon make pyrethroids ineffective. This, together with the increasing public concern about the use of toxic chemicals and their impact on the environment, has led to a flurry of interest in more environmentally acceptable insecticides and the development of more insect-tolerant cotton varieties.

Traditional breeding has done much to improve the host plant resistance of cotton to insect pests, and continues to provide new varieties which require less chemical intervention than old varieties. New characters such as glabrousness (absence of hairs on the foliage), frego bract (outward bending, thin bracts around the cotton boll), nectariless (absence of nectar-forming nectary glands on the leaves), and high gland content are now being assessed for their capacity to reduce the attractiveness of the cotton plant to insects pests. There are, however, limits to the improvement in natural resistance to insect provided by alterations in plant shape and structure. Therefore, control of insect pests in cotton cultivation depends mainly on the use of chemical insecticides that are under serious public debate for reasons of human safety and environmental pollution. Scientists have been looking for new strategies to control cotton insect pests. An attractive alternative is the production of proteins with insecticidal activity by the cotton plant itself. Genetic engineering should enhance the capacity to produce more tolerant plants by accessing a much wider gene pool for novel insect resistance characters not present in any of the *Gossypium* species or their close relatives.

Numerous laboratory and field tests confirm that the most efficient and cheapest method for protecting cotton from pests is the utilization of transgenic cotton for insect resistance. *Bt* toxins are insecticidal proteins found as parasporal crystalline inclusions in sproulated *Bt* strains. They are characterized by their potency and specificity towards specific insect pests, many of which are agronomically important, due to their relative safety to non-target insect species and vertebrates, particularly humans.

A large number of *Bt* genes have been identified, cloned and characterized, and a few have been expressed in transgenic tobacco or tomato plants where they have increased tolerance to insect pests. Scientists obtained transgenic *Bt* cotton carrying *cryIA (b)* and *cryIA (c)* genes from *Bt* spp. *kurstaki* strains HD-1 and HD-73, respectively. Although these transgenic cotton plants containing truncated *Bt* genes showed some measure of protection against insect predation in laboratory test, it soon became clear that the expression level of unmodified *cryIA* genes was too low to confer sufficient protection against cotton pests under field conditions. This is because *Bt cry* genes are typical bacterial genes in that they have a high A/T-content compared with plant genes (typically values are 60-70% for *Bt* genes and 40-50% for plant genes). As a consequence, *cry* gene codon usage is inefficient in cotton plants, and the A/T-rich regions may contain transcription germination (polyadenylation) sites (AATAAA and variations thereof), cryptic mRNA splice sites, and mRNA instability motifs (ATTTA). Scientists began modification of these *Bt* genes and since then, tremendous progress has made in this field. The effects of different degrees of gene modification were investigated in the *cryIA* genes. The results indicated that removal of the polyadenylation sites and ATTTA sequences, and changes to a total of 353 of the 615 codons, raised the levels still higher (up to 0.2-0.3% of total soluble protein) -100-fold higher than the level for unmodified genes.

Modified *Bt* genes for a number of different *Bt* toxins have been transformed into cotton plants by *Agrobacterium tumefaciens* – mediated, pollen-mediated or by the method of pathway of the pollen tube. Field performance of transgenic cotton containing *Bt* genes has shown significant caterpillar resistance. Incorporation of *Bt* genes into major commercial cultivars is expected to reduce insecticide applications for lepidopteran pests by more than 60-70%. The resistance was steadily inherited. The character was controlled by a pair of dominant genes, and was not affected by cytoplasm.

Apart from *Bt* genes, other genes for insect resistance such as those for proteinase inhibitors, a-amylase inhibitor, and lectins are also being used to produce transgenic insect-resistant cotton plants. Among these, lectins and plant proteinase inhibitors are of particular interest because they are a part of the plant's natural defence system against insect predation. Insects use many hydrolytic enzymes such as chymotrypsin or trypsin present in their gut for digestion of food protein. Earlier studies on the effects of the inhibitors and lectins induced by the diets of insects showed their effectiveness against these pests.

The scientists cloned cowpea trypsin inhibitor (*CpTI*) gene which was successfully engineered into cotton plants by *A. tumefaciens*-mediated transformation and the pollen tube pathway method. The molecular data confirmed the stability of this gene and transgenic plants had increased resistance to cotton bollworm. Cotton hypocotyl segments were transformed via the *Agrobacterium*-mediated method with soybean Kunitz trypsin inhibitor (*SKTI*) gene, and obtained transgenic cotton plants; molecular data indicated that this gene inserted stability into the cotton chromosome. The results of bioassay demonstrated that the transgenic plants showed significant resistance to the larvae of cotton bollworm.

Lectins are carbohydrate-binding proteins, some of which are toxic to insects. Various lectins have shown some toxic activity against species of the insect orders Homoptera, Coleoptera, Lepidoptera and Diptera. Recent interest has mainly concentrated on the lectin from snowdrop (GNA), because it has shown activity against aphids, which are the second most important pests of cotton. GNA gene was transformed into cotton plants using the *Agrobacterium*-mediated method. Transgenic cotton plants carrying the pea lectin (P-Lec) gene have been obtained which showed some resistance to cotton bollworm.

Although *Bt* cotton appears to be extremely effective in controlling *Lepidopteran* larvae like cotton bollworm and armyworm, there is considerable concern among the scientific and environmental groups that a single *Bt* gene is not sufficient for the long-term protection of cotton. When used like a chemical insecticide, *Bt* toxins rarely invoke resistance mechanisms in insect pests because of their short field life, which is mainly a consequence of rapid degradation by ultraviolet light. When expressed in a transgenic cotton plant, however, the insects will be constantly exposed to the toxin and this would favour the selection of resistant individuals. In order to cope with this potentially serious problem, scientists have been trying to transform two different insect-resistance genes into the same cotton plant. Cotton hypocotyls segments were infected using *A. tumefaciens* with pBinLK carrying two insecticidal genes, *P-Lec* gene and *SKTI* gene, and successfully transferred the genes into upland cotton cultivars via *Agrobacterium*-mediated transformation. Transgenic plants showed significant resistance to larvae of cotton bollworm. The plant expression vector pGB1121S4ABC was successfully transformed, harbouring both the synthesized *Bt* insecticidal protein gene and the modified *CpTI* gene, into the elite cotton cultivars.

Considerable work is directed towards targeted expression of the Bt. gene so that the gene express only in the tissue that is economically damaged by the target insect e.g. pollen, boll wall, etc. Monsanto scientists have modified the triplet codes of a Bt. gene to make it more compatible with the genetic codes preferred by the plant protein synthesizing system. Strain of *B. thuriengienesis* have been identified which are active against the Colorado potato bettle and nematodes. Hence, modified genes of the Bt. toxin will be available to combat a range of insects. More recently array of neuropeptides acting on the neural system have been identified. They have low molecular weight and effective at low quantities.

Commercialization of Bt cotton

The development of transgenic cotton that expresses *CryIA* insecticidal proteins from *Bt* spp. *kurstaki* has resulted in new varieties or lines with improved resistance to key lepidopteran insect pests. Cotton plants expressing modified *CryIA* gene sequences have demonstrated excellent control of pests such as bollworm, tobacco budworm and pink bollworm in greenhouse and field experiments. Meanwhile, transgenic *Bt* cotton did not affect the natural enemies. Field experiments showed that the total labour for pest control workdays could be decreased by 57% by planting *Bt* transgenic cotton varieties. In addition, the introduction of commercial cotton varieties producing *CryIA* insecticidal proteins is expected to reduce environmental pollution from synthetic insecticides, increase worker safety, and improve grower profitability. Thus, breeders and farmers have more interest in the breeding and commercialization of transgenic *Bt* cotton.

Once the *Bt* gene was inserted into the elite developed cotton varieties, scientists embarked on a series of tests to demonstrate the usability of the genetically modified cotton. Two methods were selected for breeding *Bt* cotton. First, *Bt* genes were directly inserted into elite cotton varieties by pollen pathway method or *Agrobacterium*-mediated transformation method. The second method is cross breeding. Once transgenic *Bt* cotton plants were obtained, scientists undertook a cross and back-cross programme to introduce the *Bt* toxin genes into genotypes of the current major varieties.

While breeding regular transgenic *Bt* cotton, scientists are at the same time developing transgenic hybrid *Bt* cotton. Compared with regular *Bt* cotton, transgenic hybrid *Bt* cotton has many fine characteristics, such as high yields, short duration for obtaining a variety, etc. Thus, breeding of transgenic hybrid *Bt* cotton was paid more attention than transgenic.

Hybrid *Bt* cotton varieties have been bred and commercialized. Field experiments showed that performance of transgenic *Bt* cotton varieties was the same as other cotton varieties except for high resistance to bollworm and budworm, and an average of 70% less chemical insecticide used than conventional cotton varieties. Transgenic hybrid *Bt* cotton varieties can increase yield 10-20% over regular *Bt* cotton or conventional cotton varieties.

Transgenic cotton for herbicide resistance

Another great progress of cotton genetic engineering is the breeding of transgenic cotton for resistance to herbicide. To impart resistance to broad spectrum herbicide has been one of the accomplishments of genetic engineering of plants. In USA, Calgene Dupont, Monsento and Phytogen are developing herbicide resistant cotton lines and some companies have gene that impart glyphosate resistance to cotton. The main motivation in developing cotton which is glyphosate resistant is its broad spectrum activity and being environmentally safe. Glyphosate is a simple chemical and degrades rapidly in the environment, precluding residual effect or possibility of pollution of ground water. Coker 312 is used as a recipient genotype. Calgene is developing cotton resistant to bromoxynil, another broad spectrum herbicide. The gene was isolated from *Klebsiella ozeanae* and codes for nitrilase. It removes nitrile atom from the compound and detoxifies it. Phytogen has isolated the cotton ALS gene and used directed mutation at the critical codon to impart resistance. Cotton is highly susceptible to the sulfonylurea class of herbicides. Dupont hopes to release a new 'super' herbicide along with the engineered resistant cotton. A gene obtained from *Alcaligenes* when introduced in cotton is shown to impart tolerance to 2, 4-D. 2,4-dichlorophenoxy-acetic acid (2, 4-D) is a popular herbicide that is useful in controlling weeds. Currently, 2, 4-D is little used in the cotton industry because cotton is notoriously sensitive to 2, 4-D and was easily damaged by drift of 2, 4-D. An increase in the tolerance of cotton to 2, 4-D would minimize spray drift damage, and may even allow the application of 2, 4-D directly onto cotton for the control of serious broad-leaf weeds. Scientists have adopted a programme to engineer a herbicide detoxification pathway in cotton using a degradative gene isolated from a bacterium. Scientists have succeeded in imparting 2, 4-D tolerance to cotton using a gene (*tfdA*) encoding 2, 4-D monooxygenase, which was obtained from the bacterium *Alcaligenes eutrophus*, and have obtained transgenic cotton resistance to 2, 4-D. The progeny were subjected to a drenching spray with 800 ppm solution of 2, 4-D and gave no indication of injury. Transformants containing *tfdA* gene exhibited 50- to 100-fold greater tolerance to 2, 4-D compared with untransformed controls. This indicated that these new constructs should ensure that the resistance gene will be highly effective in protecting cotton from damage by 2, 4-D, whether it is encountered accidentally as spray drift or from deliberate application by the farmer to control weeds. From the progenies of these transgenic cotton plants, scientists have bred some fine lines with high-yielding and good agronomic performance.

The genes for resistant herbicide bialapho and bromoxynil have been isolated, and are being transformed into cotton by the *Agrobacterium*-mediated method.

Transgenic cotton for disease resistance

Diseases are another important factor which cause huge yield loss. Two of the major diseases of cotton are *Fusarium* and *Verticillum* wilt. Some resistant genes have been isolated, and are being transformed into cotton plants.

Plants ward off pathogen infections by eliciting an array of defence mechanisms, including reinforcement of the wall, synthesis of phytoalexins and oxidation of phenolic compounds, activation of defence-related genes and localization of cell death or the hypersensitive response. Associated with these reactions is a rapid and transient production of active oxygen species (AOS), such as the superoxide anion radical (O_2^-), hydroxyl radical (OH^-), and hydrogen peroxide (H_2O_2). The accumulation of H_2O_2 and related AOS has been determined to be one of the earliest events that occurs at the host-pathogen recognition; it has been postulated to play an important role in plant defence. In addition to its oxidative potential in killing or inhibiting the growth of pathogens, H_2O_2 has been shown to be involved in a number of plant defence response processes. Glucose oxidase (GO) catalyses the oxidation of a-D-glucose by molecular oxygen, yielding gluconic acid and H_2O_2. A number of bacteria and fungi produce GO, which has a putative antibiotic function, but GO has not been found in plants. The GO gene from *Talaromyces flavus* can inhibit the growth of *Fusarium* and *Verticillum* wilt. GO genes from*Talaromyces flavus* were transformed.

Environmental stress genes

Responses of plants to stress are complex and dependent on the expression of many genes. Two stresses, drought and temperature extremes, are being investigated indepth in cotton. Search for genes whose expression is altered in drought-stressed plants, is on. A cDNA library from drought-stressed cotton has been obtained which is screened for uniquely expressed mRNA species. The influence of over-production of proline as a mechanism for stress resistance is examined. Cotton is being transformed with the bacterial "osmogene" under the regulation of a constitutive promoter. The role of superoxide dismutase in stress response is being examined like screening for the gene coding for ADH from a cDNA library made to combat anaerobic respiration during flooding.

Hybrid production

Superior hybrid cultivars are being produced having high expression of heterosis to improve yield.

Cotton is self-compatible and hence cytoplasmic male sterile lines are being raised to produce hybrid seeds. TA-29 specific promoter is being used for male sterility induction in cotton.

A concept for inducing male sterility and fertility for hybrid seed production has been developed. At the biotechnical level, the A line is transformed with a vector that carries genes which confer herbicide resistance and induce pollen sterility. A constitutive promoter is used to control expression of the resistance gene, but for sterility, an anther-specific promoter is used to control a gene that codes for ribonuclease. The ribonuclease protein degrades the RNA that is normally produced by the plant's genes for pollen development. Because proteins essential for pollen development cannot be produced as a result of the RNA destruction, the plant is male sterile. The second parent, or B line, is transformed with the same constitutively expressed herbicide resistance gene and a gene under anther-specific control that produces the antisense copy of the ribonuclease gene. In the A and B cross, either DNA-RNA or RNA-RNA sense-antisense hybridization occurs so that the ribonuclease is not produced, and the pollen develops normally.

Product quality : modifying cotton fibers

Fiber quality is determined by several measured physical and mechanical properties of mature cotton fiber. Basically, these properties are determined by fiber structure and biochemistry. In fact for every bale of cotton as many as seven quality measurements are considered. Fiber length, strength, fineness, non-fiber content, and length uniformity are measured by an electronic high volume instrument (HVI). Trained specialists evaluate colour and grade visually and also measure with HVI colorimeter. Each of the physical characteristic is determined by complex, undetermined relationship between many biologically determined factors. For instance, fiber strength, is correlated with the molecular weight distribution of cellulose polymers, the orientation of cellulose microfibrils, the frequency and distribution of reversals, and properties of fiber surfaces.

Very scanty information is available on how these strength-related parameters are biologically regulated. The precise knowledge on the initiation and cessation of cellulose biosynthesis giving rise to polymeric products with a fairly narrow molecular weight distribution is lacking. Clearly, the ultimate strength of a fiber is a composite expression of many genes. Funamental questions relating to cellulose synthesis in plant cells may improve fiber quality, but the role of other processes involving genes are also visualized. Reversal points are the spots where breakage occurs : why reversal occurs and what determines their optimal number are still unresolved questions! Inter-fiber interactions must be determining the bundle strength of the fibers and so information on fiber cuticle to identify a relationship between fatty acid chain length or degree of saturation or modification with fiber friction potential and fiber strength is needed. Similarly, genetic background of genotype is crucial for determining the fiber length. Several processes linked with turgor driven cell expansion are involved and possibly regulate gene products including malate formation and vascular proton ATPases. Wall hydrolases and peroxidases appear to be other enzymes concerned with the determination of fiber length. Generally, longer fibers tend to be fine fiber, and the longer fibers result from extended period of fiber elongation or high rate of fiber elongation. Precisely no information is available on the regulatory triggers which enforce cessation of fiber elongation and initiation of secondary wall thickening.

Cotton biotechnology involves several programmes : identification of genes which confer desirable traits; identification of a system for the stable and optimal introduction of new genes; and recovery of transgenic plants via regeneration or meristem culture.

The first aspect concerning identification of new genes for fiber improvement has lagged behind though success has been achieved in obtaining stable regeneration of cotton from embryogenic suspension cultures. *Agrobacterium*-mediated transformation of cotton suspention cultures with marker genes and then recover mature plants, introducing *Bacillus thuringiensis* toxin genes for pest resistance and also various herbicide resistance genes have been manipulated in cotton. Cotton has been transformed via particle bombardment. The introduction of resistance genes may affect fiber quality and yield.

Two complimentary approaches could be undertaken for the identification of new genes that will improve cotton fiber quality. The first, concerns the isolation and characterization of genes which are suspected to be involved in some aspect of fiber growth e.g. cellulose synthesis, the fiber cytoskeleton, wall-associated proteins, cuticle lipid synthesis, turgor generating metabolic processes, identifying genes that are regulated by phytohormones at the time of differentiation, ABA biosynthesis and IAA synthesis expressed differential during a cool temperature event; identification of key components in the fiber during deposition notably modified by cool temperature to isolate the corresponding genes and fuse these with an appropriate promoter from another species, which is more tolerant to cool temperature, to promote sugar transport to the fiber under cool temperature regimes. Most of these identifications are being attempted through quantitative comparisons of mRNA and/or protein product levels in cotton varieties differing in physical fiber traits; heterologous probes could be used for the identification of cellulose synthase genes in cotton. The second approach concerns identification of new fiber quality related genes through genome mapping technique using RFLP or PCR methods.

Based on accumulated knowledge of the biochemistry of fiber formation and development, an investigation of the genes critical to fiber differentiation and development has begun. Apart from these, some genes relevant to fiber strength and length have been sought and cloned. Rabbit hair keratin gene has been cloned and transformed successfully into cotton plants. The length of the transgenic cotton fiber is 60% higher than normal cotton fibers, and their elasticity and thermal preservation are better than normal cotton fiber.

The cellulose-synthase operon from *Acetobacter xylinum* has been clones in cotton and wrinkle-resistance cotton has already hit the USA market. Similarly, naturally coloured cottons are no more fantasies. Cotton with natural pink, cream and blue colour are a reality. It is expected that among the genetically engineered crops, cotton will be one of the earliest crops to reach the market. Of course, release with a single additional gene which confers insecticidal or herbicidal properties has been made. Cultivars having multiple genes, or perhaps whole new pathways, bestowed with improved fiber traits or stress resistance will be available. Concomitantly chemical companies are developing new, more eco-friendly herbicides that would be effective at low doses compared with those currently being used. The chief aim of exploiting new biotechnological techniques is to improve quality parameters e.g. strength, length and micronaire in transgenic plants of cotton.

In summary, added information on physiology of fiber development and availability of biotechnological tools would provide opportunities to improve cotton fiber higherto not visualized. Genomic mapping and exploitation of targeted approach of gene identification and transfer should help in achieving better cottons.

□□□

14

Fiber Quality of Transgenic Cotton

Cotton varieties containing traits inserted by recombinant DNA technology ("transgenic" cotton) contain novel input characters that allow plants to be resistant to insects (e.g., Bt cotton) or tolerance to certain herbicides [e.g., Roundup Ready (RR) and BXN cotton). Some varieties are "stacked", i.e., individual plants exhibit both insect resistant and herbicide tolerance (Bt/RR). About 30 per cent of the world production of cotton in previous years was transgenic varieties grown in Argentina, Australia, China, India, Indonesia, Mexico, South Africa and the USA. A second wave of transgenic cottons with improved quality are being introduced and/or are in the pipeline to be available over the coming years. The popularity of transgenic cotton varieties is due to improvements in insect pest management and additional options in grass and weed control. A review of safety information indicates that transgenic cotton does not pose any different risks to human or animal health than conventional cotton and there are no adverse effects on the environment. Agronomic and cultural benefits are well accepted. However, there have been questions raised as to the effect of transgenic cotton technology on fiber quality. There is much misinformation based on anecdotal information and perception rather than science. Commercial pressures provide more than significant incentives for breeders to maintain quality equivalency in new cultivars. Plant breeders consider yield, input traits and fiber quality as high priority breeding objectives.

Anecdotal reports and opinions of the effect of biotechniques on lint quality are mixed and oftentimes confusing. Evaluation of the question of whether biotechnology has had an effect on fiber quality is made complex because of the masking environmental, management, and weather effects (i.e., climate, soils, agronomic practices and location); non-statistically valid comparisons and comparisons that do not take these parameters into account are unreliable. The most effective method to minimize the random effects of climate, soils and location is to grow varieties side by side under the same field conditions.

Lint Quality

The extensive processing studies comparing transgenic variants with their recurrent parents have been conducted. High Volume Instrumentation (HVI) results of micronaire, leaf grade, yellowness, length (UHML) uniformity, strength and elongation showed no main effect differences due to transgenic technology. The only lint quality parameter that showed a statistically significant difference related to transgenic technology was reflectance; there was a slight increase in reflectance among transgenic varieties as compared with conventional. Even though effect for reflectance is statistically significant, the small difference is not of practical importance from a textile point of view. Advanced Fiber Information System (AFIS) multidata showed no main effect differences when length, short fiber content, upper quartile length, maturity ratio and 12 other evaluations were considered. Shirley Analyzer evaluations of non-lint content showed no significant differences due to transgenic technologies. Stelometer-strength and elongation percent were not different either.

Yarn characteristic evaluations of ring and rotor spun yarns showed no effect of practical significance of the variant types for all variables and yarn counts; hence, these tests showed that yarns produced from RR and Bt variants were not different from the original varieties. In all fabric characteristic evaluations, including greige state and dyeing, the RR and Bt variants looked at least equal to the original variety.

In an economic evaluation, cotton varieties were compared for commercial yields, and quality. The entomological and economic impact of Bt cotton when compared with conventional cotton, no meaningful differences in grade and staple length between transgenic and conventional varieties was observed. Micronaire, fiber strength, fiber length and fiber length uniformity were not affected, demonstrating that use of glyphosate (Roundup Ready) did not have a detrimental effect on fiber quality. Conventional varieties exhibited slight advantages in mean length and uniformity and transgenic varieties were slightly better (lower) in micronaire. Conventional varieties in tests had about 4 per cent higher strength. While glyphosate tolerant (Roundup Ready, EE) types are the tolerant cotton varieties which continue to be an important option for some growers. The transgenic BXN was slightly longer, finer (lower micronaire) and higher in strength than the recurrent parent, though differences were slight. The most reliable comparisons for evaluating potential effects on quality are those in which each unique transgenic variant is paired with its original conventional recurrent parents grown under the same year, location, environment, and management regime, plants are thus subjected to the same biotic and abiotic stresses. A head to head comparison of fiber qualities provides statistical rigor for a valid comparison.

Strength : The comparisons shown revealed that the means of fiber strength of Bt variants is about 1.1 per cent less than the means of the corresponding recurrent conventional parents. Roundup Ready (RR) varieties show a slight improvement of strength as did the stacked Bt/RR lines. The BXN varieties showed no difference from the recurrent parent.

Staple length : The length reported was also only slightly different. Bt lines showed no difference in mean staple length from the recurrent parents whereas staple length of BXN was slightly better. Length of RR and Bt/RR was less than one per cent below the mean length of the parental lines

Micronaire : The micronaire value changes were most dramatic of all fiber parameters. The mean micronaire value for 535 parental entries was 4.6 whereas the mean of all transgenic lines is about 4.5. Bt variants had micronaire values nearly 3 per cent lower than values of parental lines. All transgenic lines showed some reductions in average micronaire. Many view the slightly lower trend for micronaire a desirable factor especially for years in which conditions may be favourable to produce high-micronaire fibers sufficient to become discounted in value. It was suggested that a slightly lower micronaire value could mean more immature fibre.

Length Uniformity : The differences in length uniformity were not dramatic. The maximum difference was with Bt lines in which the mean Length Uniformity Index decreased by only 0.3 per cent. RR varieties decreased 0.2 per cent and stacked Bt/RR and BXN varieties were the same as their parents.

Yield : A look at yield differences with recurrent parents may suggest a partial explanation as to why some of the fiber quality parameters differ. There were consistent yield advantages for transgenic variants containing Bt, whether Bt alone or stacked Bt/RR. Most experts believe the yield increase can be explained by the reduced yield loss due to benefits of sub-threshold (i.e., low worm pressures, which do not justify application of insect treatments) control of damaging insects. With the Bt gene ever present in the plant, insect control at even the lowest levels of worm pressure is provided. If this theory is correct, then likely there are additional cotton bolls protected that typically would have been sacrificed to insects, thus, either lengthening the growing season or intensifying the existing boll retention period.

□□□

Index

— P

— R —

— S —

□□□